Spatial Perception and Navigation in the Absence of Vision

Spatial Perception and Navigation in the Absence of Vision

Guest Editors

Daniel-Robert Chebat
Maurice Ptito

MDPI

Basel • Beijing • Wuhan • Barcelona • Belgrade • Novi Sad • Cluj • Manchester

Guest Editors

Daniel-Robert Chebat
Department of Psychology
Ariel University
Ariel
Israel

Maurice Ptito
École d'Optométrie
Université de Montréal
Montreal
Canada

Editorial Office
MDPI AG
Grosspeteranlage 5
4052 Basel, Switzerland

This is a reprint of the Special Issue, published open access by the journal *Sensors* (ISSN 1424-8220), freely accessible at: https://www.mdpi.com/journal/sensors/special_issues/Spatial_Perception_and_Navigation_in_the_Absence_of_Vision.

For citation purposes, cite each article independently as indicated on the article page online and as indicated below:

Lastname, A.A.; Lastname, B.B. Article Title. *Journal Name* **Year**, *Volume Number*, Page Range.

ISBN 978-3-7258-7236-7 (Hbk)
ISBN 978-3-7258-7237-4 (PDF)
https://doi.org/10.3390/books978-3-7258-7237-4

Contents

About the Editors

Daniel-Robert Chebat

Daniel-Robert Chebat is a cognitive and systems neuroscientist whose research focuses on brain plasticity, perceptual learning, multisensory processing, and spatial navigation, with a particular emphasis on sensory deprivation and blindness. He directs the Visual and Cognitive Neuroscience Laboratory (VCN Lab) and serves as Scientific Director of the Navigation and Accessibility Research Center (NARCA) at Ariel University.

A central feature of his work is the use of behavioral learning paradigms—especially maze-based navigation tasks—to investigate how humans acquire, refine, and generalize spatial knowledge under altered or constrained sensory conditions. These paradigms are designed to capture learning dynamics, strategy formation, and adaptation over time rather than static performance.

His research demonstrates that navigation learning can occur independently of visual experience and that cortical systems traditionally considered "visual" can support non-visual perceptual learning and spatial computations. By combining controlled behavioral paradigms with neuroimaging methods (fMRI, structural MRI, and connectivity analyses), his work highlights the task-selective and amodal organization of the human brain.

A major contribution of this program is establishing sensory substitution as a powerful model for studying perceptual learning and neuroplasticity, with direct implications for rehabilitation, accessibility technologies, and assistive system design.

Maurice Ptito

Maurice Ptito is a Professor of Visual Neuroscience at the School of Optometry at Université de Montréal and an internationally recognized authority on visual system development, neuroplasticity, and sensory substitution. He is also an adjunct professor at the Montreal Neurological Institute (McGill University) and a guest scientist at the University of Copenhagen, reflecting a long-standing record of international collaboration.

Prof. Ptito's research focuses on how the human brain adapts to visual deprivation, both congenital and acquired, using a multidisciplinary approach that integrates neuroimaging (fMRI, PET, DTI), electrophysiology, neuromodulation, and behavioral methods. He is particularly known for his pioneering contributions to sensory substitution research, including seminal work on the Tongue Display Unit (TDU), which demonstrated that tactile input can recruit visual cortical areas in blind individuals, providing compelling evidence for cross-modal plasticity.

Over the course of his career, Prof. Ptito has authored or co-authored more than 220 peer-reviewed publications, book chapters, and conference papers spanning neuroscience, vision science, brain imaging, and rehabilitation. His contributions have been widely recognized through numerous honors, including Fellowship in the Royal Society of Canada and appointment as a Knight of the National Order of Québec. Prof. Ptito continues to shape the field of visual neuroscience.

sensors

MDPI

Editorial

Spatial Perception and Navigation in the Absence of Vision

Daniel-Robert Chebat [1,*] and Maurice Ptito [2,3,4]

1 Visual and Cognitive Neuroscience Laboratory (VCN Lab), The Department of Psychology, Ariel University, Ariel 40700, Israel
2 Department of Neurology and Neurosurgery, Ecole d'Optométrie, Université de Montréal, Montreal, QC H3T 1J4, Canada; maurice.ptito@umontreal.ca
3 Montreal Neurological Institute, McGill University, Montreal, QC H3A 2B4, Canada
4 Department of Applied Mathematics and Computer ScienceVisual Computing, Danish Technical University, 2800 Kongens Lyngby, Denmark
* Correspondence: danielc@ariel.ac.il

In congenital blindness (CB), tactile and auditory information can be reinterpreted by the brain to compensate for visual information through brain plasticity mechanisms triggered by training. Visual deprivation may not cause a cognitive spatial deficit since blind people can acquire spatial knowledge about the environment. This spatial competence takes longer to achieve but is eventually reached through training-induced plasticity. Even though complete visual deprivation leads to volumetric reductions in brain structures associated with spatial learning, blind individuals are still able to navigate. However, the neural structures involved in this function are not fully understood. In this Special Issue, leading experts in the field submitted reviews (2) or original research (8) on spatial navigation in the absence of vision, using a variety of approaches to advance our understanding of multisensory spatial knowledge acquisition. Our aim is for these topics to be explored in people who are congenitally blind, late blind, and sighted who use multisensory cues to perceive spatial information, as well as discussing their abilities, strategies, and corresponding mental representations. We therefore organized the chapters into three cohesive themes to offer readers a structured exploration of the multifaceted dimensions of spatial perception and navigation in the absence of vision. These sections allow for a comprehensive exploration, delving into the understanding of spatial navigation, innovative assistive technologies, and the intricacies of multisensory experience and spatial cognition.

Theme 1: Understanding Spatial Navigation in the Absence of Vision.

Heled, Elul, Ptito, and Chebat delve into the intricate landscape of deductive reasoning and working memory skills within individuals living with blindness. These cognitive functions, vital for executive functioning in everyday life, play a pivotal role in shaping the thought processes and actions of individuals, even more so for those who are visually impaired. The study endeavors to shine a light on the interplay between visual experience, deductive reasoning abilities, and working memory skills, offering a nuanced perspective of cognitive function in the absence of sight. Moreover, it challenges preconceived notions about the role of visual experience in cognitive development and underscores the profound interdependence of deductive reasoning and working memory skills within the visually impaired population. This nuanced exploration serves as a testament to the adaptability and resilience of the human mind, shedding light on the incredible potential for cognitive growth and innovation in individuals who navigate the world without the benefit of sight. In their groundbreaking study, *Real and Araujo* explore the impact of network Quality of Service (QoS) on spatial perception through sensory substitution in navigation systems designed for individuals who are blind and visually impaired. Their

https://doi.org/10.3390/s25030960

research underscores the delicate balance between the acquisition of spatial information and temporal constraints in the rapidly evolving landscape of assistive technologies. By conducting tests with participants under varying delay conditions between motor actions and triggered stimuli, the study unveils a trade-off between spatial information acquisition and performance degradation. Moreover, it sheds light on a learning curve in scenarios marked by impaired sensorimotor coupling. In summary, Real and Araujo's research advances our understanding of how network QoS influences spatial perception through sensory substitution in navigation systems.

Theme 2: Innovative Assistive Technologies for Spatial Perception.

Kilian, Neugebauer, Scherffig, and Wahl introduce a groundbreaking innovation, the "Unfolding Space Glove", an open-source sensory substitution device. This remarkable device transforms hand gestures into vibratory stimuli, enabling individuals with blindness to explore and navigate their spatial environment haptically. By providing real-time feedback on the relative position and distance of nearby objects through vibrations on the back of the hand, the Unfolding Space Glove empowers blind users in tasks such as object recognition and wayfinding. What makes this device truly revolutionary is its portability, capacity to operate in diverse lighting conditions, its provision of immediate feedback, and its unobtrusive design, rendering it a promising tool for enhancing mobility and spatial awareness among individuals with visual impairments. In the next paper, *Osiński, Łukowska, Hjelme, and Wierzchoń* present the innovative "Colorophone 2.0", a wearable color sonification device. This remarkable device has the potential to translate spatial color information into real-time stereo soundscapes. The primary objective is to bridge the sensory gap for individuals with visual impairments, providing them with the ability to perceive and interact with color-rich environments through auditory means. The authors delve into the design and development of this sensory substitution device and present a usability evaluation, highlighting its potential to empower those who cannot directly access visual information. The study by *Bleau, Paré, Djerourou, Chebat, Kupers, and Ptito* focuses on evaluating the EyeCane, an electronic travel aid, for its effectiveness in helping individuals with visual impairments navigate obstacles. Vision loss significantly impacts a person's ability to move around safely. Various devices have been developed to assist the visually impaired, and the EyeCane is an inexpensive and easy-to-use ETA designed to detect obstacles within a 2 m range. Participants underwent training and testing using the EyeCane in a life-sized obstacle course containing different types of obstacles, such as cubes, door frames, steps, and posts. All subjects quickly learned to use the EyeCane and successfully completed the trials. Early blind participants were the fastest to navigate the obstacle course, followed by late blind and sighted controls. The EyeCane shows promise as a primary mobility aid for blind navigation, but there are challenges related to its ability to reliably detect lower obstacles on the ground.

Theme 3: Multisensory Experience and Spatial Cognition.

Afonso-Jaco and Katz's review highlights the transformative potential of auditory exploration in individuals who are blind. They investigate the role of auditory information and virtual reality in the development of spatial cognition for blind individuals. The paper underscores the importance of using spatial audio technology and virtual reality as tools to study higher-level cognitive processes related to spatial cognition in the absence of vision. It also touches upon the influence of early visual experience on spatial abilities among the blind, highlighting the complexity of this topic. The authors provide insights into how blind individuals acquire spatial information, create mental representations of their surroundings, and navigate their environment. This research offers valuable contributions to the field of spatial cognition and holds promise for enhancing the spatial abilities of the visually impaired.

In the next paper, *Lahav* immerses us in the world of virtual reality as an orientation aid for individuals who are blind. Their research investigates the impact of virtual environments on the exploration process, cognitive map construction, and orientation tasks, paving the way for immersive and inclusive navigation solutions. The research focused on comparing interactions with different systems, namely BlindAid, Virtual Cane, and real spaces, to assess their influence on the exploration process, the construction of cognitive maps, and the ability to perform orientation tasks. The results demonstrated that participants could effectively explore unfamiliar spaces, construct cognitive maps, and perform orientation tasks when using both virtual systems and real spaces. This paper underscores the potential of virtual reality systems in aiding the spatial orientation of individuals who are blind. It highlights the value of unique user–interface commands, which enhance exploration, cognitive map construction, and orientation performance. Muteness is a prevalent disability with various technological solutions aimed at transforming mute languages into vocal speech. The study by *Holdengreber, Yozevitch and Khavkin*, on an Intuitive Cognition-Based Method for Generating Speech Using Hand Gestures, introduces a novel approach that does not require prior sign language knowledge but instead leverages intuitive cognition to generate speech through hand gestures. This study presents a promising alternative for individuals with speech disabilities, offering an intuitive and adaptive speech generation method. Future enhancements could include real-time adaptive filtering and expanded phoneme recognition to improve fluency and accuracy. Finally, the study by *Leah Fostick and Nir Fink* explores how various hearing protection devices (HPDs) impact sound localization, with a focus on three distinct stimuli: spoken words, pink noise, and gunshots. The findings unearth intriguing insights into the complex interplay between sound source, hearing protection, and the listener's auditory prowess. They highlight the complexity of sound localization and its relationship with hearing protection, offering insights that resonate with both researchers and individuals with hearing impairments.

As Guest Editors, we extend our heartfelt gratitude to each contributor for their dedication to advancing our understanding of spatial perception and navigation in the absence of vision. These insights not only broaden the horizons of scientific knowledge but also hold promise regarding their ability to enrich the lives of individuals with blindness or visual impairments.

Conflicts of Interest: The authors declare no conflict of interest.

List of Contributions:

1. Bleau, M.; Paré, S.; Djerourou, I.; Chebat, D.; Kupers, R.; Ptito, M. Blindness and the Reliability of Downwards Sensors to Avoid Obstacles: A Study with the EyeCane. *Sensors* **2021**, *21*, 2700. https://doi.org/10.3390/s21082700.
2. Holdengreber, E.; Yozevitch, R.; Khavkin, V. Intuitive Cognition-Based Method for Generating Speech Using Hand Gestures. *Sensors* **2021**, *21*, 5291. https://doi.org/10.3390/s21165291.
3. Fostick, L.; Fink, N. Situational Awareness: The Effect of Stimulus Type and Hearing Protection on Sound Localization. *Sensors* **2021**, *21*, 7044. https://doi.org/10.3390/s21217044.
4. Osiński, D.; Łukowska, M.; Hjelme, D.; Wierzchoń, M. Colorophone 2.0: A Wearable Color Sonification Device Generating Live Stereo-Soundscapes—Design, Implementation, and Usability Audit. *Sensors* **2021**, *21*, 7351. https://doi.org/10.3390/s21217351.
5. Lahav, O. Virtual Reality Systems as an Orientation Aid for People Who Are Blind to Acquire New Spatial Information. *Sensors* **2022**, *22*, 1307. https://doi.org/10.3390/s22041307.
6. Kilian, J.; Neugebauer, A.; Scherffig, L.; Wahl, S. The Unfolding Space Glove: A Wearable Spatio-Visual to Haptic Sensory Substitution Device for Blind People. *Sensors* **2022**, *22*, 1859. https://doi.org/10.3390/s22051859.
7. Heled, E.; Elul, N.; Ptito, M.; Chebat, D. Deductive Reasoning and Working Memory Skills in Individuals with Blindness. *Sensors* **2022**, *22*, 2062. https://doi.org/10.3390/s22052062.

8. Afonso-Jaco, A.; Katz, B. Spatial Knowledge via Auditory Information for Blind Individuals: Spatial Cognition Studies and the Use of Audio-VR. *Sensors* **2022**, *22*, 4794. https://doi.org/10.3390/s22134794.
9. Real, S.; Araujo, A. Network QoS Impact on Spatial Perception through Sensory Substitution in Navigation Systems for Blind and Visually Impaired People. *Sensors* **2023**, *23*, 3219. https://doi.org/10.3390/s23063219.

sensors

MDPI

Article

Deductive Reasoning and Working Memory Skills in Individuals with Blindness

Eyal Heled [1,2,3,*], Noa Elul [2], Maurice Ptito [4,5,6] and Daniel-Robert Chebat [1,3,7]

1 Department of Psychology, Ariel University, Ariel 40700, Israel; danielc@ariel.ac.il
2 Neurological Rehabilitation Department, Sheba Medical Center, Ramat Gan 52621, Israel; noaelul10@gmail.com
3 Navigation and Accessibility Research Center (NARCA), Ariel University, Ariel 40700, Israel
4 School of Optometry, University of Montreal, Montreal, QC H3T 1P1, Canada; maurice.ptito@umontreal.ca
5 Department of Neuroscience, Copenhagen University, 2200 Copenhagen, Denmark
6 Department of Neurology and Neurosurgery, Montreal Neurological Institute, McGill University, Montreal, QC H3A 2B4, Canada
7 Visual and Cognitive Neuroscience Laboratory (VCN Lab), Department of Psychology, Faculty of Social Sciences and Humanities, Ariel University, Ariel 40700, Israel
* Correspondence: eyalheled@gmail.com; Tel.: +972-502-310313; Fax: +972-3-9191437

Abstract: Deductive reasoning and working memory are integral parts of executive functioning and are important skills for blind people in everyday life. Despite the importance of these skills, the influence of visual experience on reasoning and working memory skills, as well as on the relationship between these, is unknown. In this study, fifteen participants with congenital blindness (CB), fifteen with late blindness (LB), fifteen sighted blindfolded controls (SbfC), and fifteen sighted participants performed two tasks of deductive reasoning and two of working memory. We found that while the CB and LB participants did not differ in their deductive reasoning abilities, the CB group performed worse than the sighted controls, and the LB group performed better than the SbfC group. Those with CB outperformed all the other groups in both of the working memory tests. Working memory is associated with deductive reasoning in all three visually impaired groups, but not in the sighted group. These findings suggest that deductive reasoning is not a uniform skill, and that it is associated with visual impairment onset, the level of reasoning difficulty, and the degree of working memory load.

Keywords: congenital blindness; late blindness; deductive reasoning; working memory; executive functions

1. Introduction

Executive functions are essential for regulating and organizing our thoughts and actions for the purpose of goal-directed behavior in everyday life [1–3]. Working memory [1,3] and deductive reasoning [4,5] are two of the executive functions' core high-level cognitive abilities. Working memory is responsible for the active maintenance and manipulation of information in the short term [6]. Reasoning is defined as the ability to make inferences, manipulations, and alterations of information. Deductive reasoning is a specific type of higher reasoning skill using verbal, visual, or numeric premises to produce a logical conclusion [7].

Deductive reasoning and working memory are associated with one another in people who are sighted [7–11], and these skills are dependent on perceptual aspects [4], but the exact influence of visual experience on these skills remains unclear. Two opposing theories predict different cognitive performance subserved by neural plasticity in individuals with blindness. The first, the "sensory compensation hypothesis", suggests that damage in a certain modality may lead to superior performance on cognitive tasks in the intact remaining modalities due to plastic reorganizational changes [12]; the second, the "perceptual deficit hypothesis", assumes that damage to one sense leads to impairment in the other senses [13].

Research on deductive reasoning and working memory in people who are blind does not clearly support either of these [13–15]. For instance, in terms of working memory tested via the auditory modality, certain studies indicate better performance by people who are blind [16–19], while others find they perform equally well compared to sighted controls [20–22]. Working memory can also be tested via the tactile modality, and here too there exist inconsistencies in the literature, as certain studies show superior or equal performance by people with blindness compared to sighted controls [17,22–24], while others show deficiencies [22,25–27]. In terms of deductive reasoning, the same contradictions exist since certain findings indicate that blind individuals perform worst on deductive reasoning tasks compared to sighted controls [28–30], yet others indicate that they perform equally well or better than controls [31–35].

Visual deprivation from birth affects the types of strategies employed [36] and may even confer supranormal abilities to people who are CB in terms of working memory [37], making them immune to certain impedance effects of irrelevant visual information for mental imagery tasks [28]. Comparing these abilities between individuals with CB to those with late-onset blindness (LB) has yielded inconsistent findings [31,33,35,38–40]. Two factors can account for this advantage. The first stems from neurological processes at the structural and functional levels that occur as a result of congenital blindness and the second involves mechanisms of training induced brain plasticity (for a review, see [15]), suggesting that intensive use of working memory and enhanced practice leads to improvements compared to those who have not used it as much or rely on other functions for performance [23]. Furthermore, despite the fact that both working memory and deductive reasoning are crucial skills for people who are blind, this relationship is not fully understood in blindness.

This study explores differences between CB, LB, and sighted participants in terms of their working memory and reasoning skills, as well as the relationship between these two skills in groups with varying degrees of visual experience. Two different measures for working memory and for reasoning skills are used in this study: working memory is measured using both the letter–number sequencing task and the digit span backward task, while the deductive reasoning tests are used to evaluate reasoning skills via audition (see Section 2). For each group, the scores for each test and each item are calculated and the relationship between deductive reasoning and verbal working memory is assessed. Our general findings show that vision is not essential for the development of deductive reasoning or working memory skills. Importantly, we also find that deductive reasoning skills are predicted by working memory skills. Psychologists and other professionals in the field of cognitive assessment in educational and vocational settings could consider these important aspects and pay more attention to the relationship between onset of blindness and the abilities that are being measured.

2. Materials and Methods

2.1. Participants and Ethics

Sixty native Hebrew speakers (30 men), aged 20–43 (M = 29.41, SD = 6.19) and possessing between 12 and 24 (M = 14.46, SD = 2.08) years of education were included in the study. The sample is composed of sighted, CB, and LB individuals, separated into four groups. A total of 30 blind participants with no residual vision or light perception recruited from the Center for the Blind in Israel were divided into two groups: 15 CB participants (9 men) and 15 LB participants (8 men). The average age of blindness onset in the LB group was 19.66 (SD = 8.65). Additionally, 30 sighted controls with normal vision were divided into two groups: 15 blindfolded participants and 15 using full vision for the task. In order to control for the load on working memory during the reasoning task, a sighted full vision control group performed the task using a pen and paper (see Table 1 for demographics). Using a pen and paper reduces the load on working memory for this task. We excluded participants based on any of the following: (a) developmental disorders with a potential effect on cognitive functioning; (b) diagnosis of mental health disorders, either past or present; (c) past or present diagnosis of neurological disorders; and (d) non-native Hebrew speakers. We included participants: (a) above 11 years of education; (b) above 18 years of age; and (c) for sighted participants, those with intact vision. All participants provided written informed consent prior to participation in the study, which was approved by the Ethics Committee of Ariel University.

Table 1. Mean (standard deviation in parenthesis) age and education of participants by group.

	Congenitally Blind	Late Blind	Blindfolded Controls	Sighted Controls
Age	34.4 (5.76)	29.93 (4.36)	26.33 (4.36)	26.33 (4.45)
Education (years)	15.33 (2.09)	13.93 (1.33)	14.93 (2.65)	13.86 (1.55)

2.2. Materials

2.2.1. Deductive Reasoning Tests

(1) Word Context Test [41]. In this task, the meanings of 10 unfamiliar words are deduced based on five clue sentences, which provide some information about the meaning. The clue sentences are presented orally, one at a time, and contain progressively more detailed information. Participants are able to provide an answer after each sentence. If correct, he/she continues to the next word, and, if not, they are presented with another clue sentence that is more specific. The objective of the task was to correctly guess the meaning of the gibberish word using as few clue sentences as possible; the sooner the correct answer is reached, the higher the score, which ranges between 1 and 5 for each presented word. The total score of the test ranges from 0 to 50 with no time limit for this task. The dependent variable is the total correct score in the task. An example item of the task:

- Most people need to *prifa* several times a day,
- Most people are very careful about their *prifa*.

(2) Deductive reasoning argument task. This test is taken directly from the vocation assessment battery that is used for evaluating cognitive capabilities of applicants for jobs in vocational assessment centers in Israel. Vocational centers in Israel administer the test using a pen and paper for the sighted and orally for people who are blind. In this task, twenty-five reasoning arguments are presented orally. Participants have to choose the correct answer among four possible answers. One point is given for each correct answer, ranging from 0 to 24, without a time limit for the task. The score for this task is the number of correct answers. An example question:

Marcy is Dana's daughter and Bella's mother. What is Dana for Bella?

1. Her mother.
2. Her daughter.
3. Her grandmother.
4. Her granddaughter.

2.2.2. Working Memory Tests

(1) The Digit Span Backward [42]. A task that is specifically used to test the ability to manipulate information [43]. Participants are to repeat a sequence of digits presented orally, in reverse order, beginning with two digits for each sequence length and up to nine digits. If the participants are correct, another single digit is added to the sequence which can reach a total of 16 trials. This procedure is repeated until participants fail two consecutive sequences of the same length. For every correct response, the participants receive one point, within a score range of 0 to 16. The dependent variable for the task is the total number of correct recalls.

(2) Letter–Number Sequencing [42]. In this task, a series of random numbers and letters are presented orally to the participants, who are required to repeat the sequence. First, the numbers in ascending order, and then the letters in alphabetical order. The task is

stopped when the participants make an error in three consecutive trials of the same length. A potential total of 21 trials are administered, and 1 point is provided for every correct answer, within a scoring range of 0 to 21. The sum of the correct points is considered as the Letter–Number Sequencing score.

2.3. Procedure

Suitable participants according to our exclusion and inclusion criteria received a short explanation of the study, signed an informed consent form (which was read to them in Hebrew), answered the examiners' demographic questionnaire, and then performed the cognitive tasks. The experiment lasted about 60 min, and the participants were able to ask questions regarding the study after completing the study.

2.4. Statistical Analysis

The groups were compared in the demographic measures using Multivariate Analysis of Variance (MANOVA) for age and education and Chi-square analysis for gender. A one-way ANOVA was used to compare the reasoning and working memory scores of the four groups, followed by a Tukey post hoc test to clarify the source of the differences. A non-parametric Kruskal–Wallis H test was also conducted in all comparisons to account for the relatively small sample size. Next, a working memory composite score was created by averaging the two tests after transforming them to Z-scores to explore the extent to which deductive reasoning can be explained by working memory in each group. We combined the scores on their common theoretical and empirical grounds [42] and the pattern of results in the current study. The two reasoning tasks, however, reflect slightly different aspects of reasoning and so were not combined. A regression analysis was used for the working memory composite score of each of the reasoning tasks separately to extract the R-squared scores for the explained variance. The p value was adjusted for multiple comparisons of each reasoning task cluster of regressions to $p = 0.012$. We were then able to compare the correlation estimates by using Fisher's R-to-Z transformation test to assess the differences between the groups. We used SPSS software 25 to analyze the data, and the significance level for all the tests was $p < 0.05$.

3. Results

No differences were found between the groups in terms of age and years of education ($F(3,56) = 1.58$, $p = 0.132$), or gender ($\chi^2(3) = 0.13$, $p = 0.721$).

3.1. Reasoning Tasks

(1) Word context task. In the word context test, no significant effect was found between the groups ($F(3,59) = 1.12$, $p = 0.348$, $\eta_p{}^2 = 0.05$, H(3) = 3.16, $p = 0.367$; see Figure 1).

(2) Deductive reasoning argument task. Comparing the four groups in the deductive reasoning argument task yielded a significant result ($F(3,59) = 8.18$, $p < 0.001$, $\eta_p{}^2 = 0.3$, H(3) = 18.39, $p < 0.001$), showing that the sighted participants performed better than the CB group and the blindfolded group, with LB performing better than the blindfolded group.

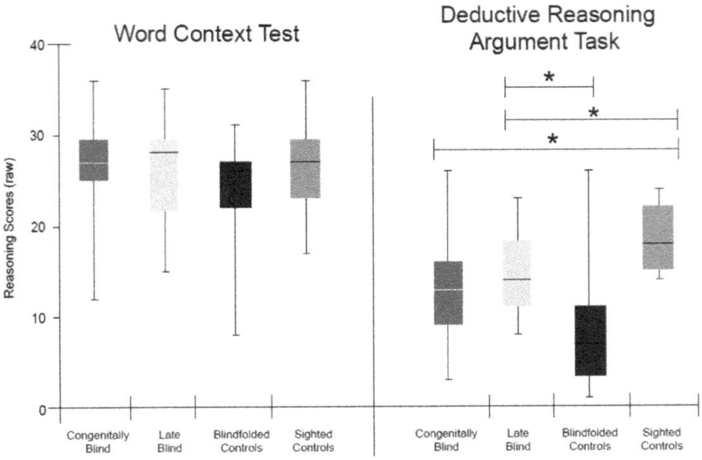

Figure 1. Group comparisons in the reasoning tests scores. Bar graph comparing the performance of congenitally blind, late blind, blindfolded, and sighted controls in the word context test (**left**) and the deductive reasoning argument task (**right**). No difference was found between the groups in the word context test. In the deducting reasoning argument task, the sighted controls performed better than the congenitally blind and blindfolded, and the late blind group performed better than the blindfolded group. * $p < 0.05$.

3.2. Memory Tasks

(1) Digit span backwards. A comparison of the groups in the digit span backwards showed a significant effect ($F(3,59) = 5.58$, $p = 0.002$, $\eta_p{}^2 = 0.23$, $H(3) = 14.99$, $p = 0.002$) where CB performed better than the rest of the groups.

(2) Letter–number sequencing task. Similarly, this task was also significant ($F(3,59) = 4.28$, $p = 0.008$, $\eta_p{}^2 = 0.18$, $H(3) = 10.39$, $p = 0.015$), showing again, using the post hoc test, that CB outperformed the other groups (see Figure 2).

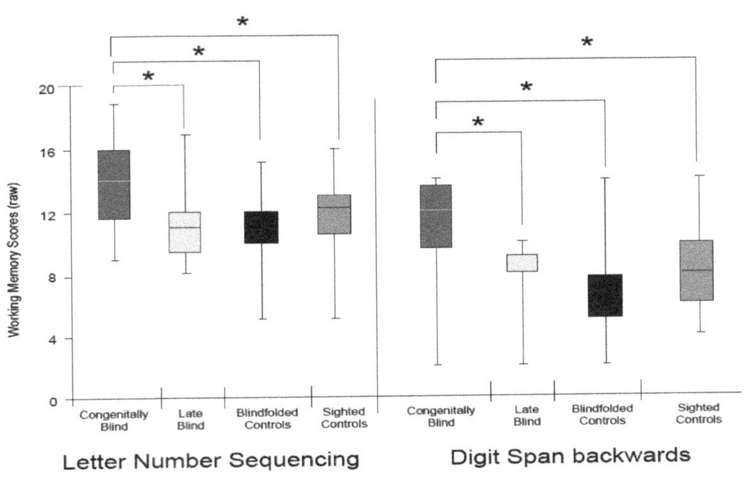

Figure 2. Group comparisons in the working memory tests scores. Bar graph comparing performance of congenitally blind, late blind, blindfolded, and sighted controls in the letter–number sequencing task (**left**) and the digit span backwards task (**right**). The congenital blind group performed better than all the other groups in both working memory tasks. * $p < 0.05$.

3.3. Relationship between the Variables

The regression analysis between the working memory composite score and the word context test is not significant in the CB ($F(1,14) = 1.35$; $p = 0.265$), LB ($F(1,14) = 1.15$, $p = 0.302$), blindfolded ($F(1,14) = 0.42$, $p = 0.527$), and sighted participants ($F(1,14) = 1.57$, $p = 0.232$). However, the regression between the working memory composite score and deductive reasoning argument task yields significant results for the CB participants ($F(1,14) = 10.66$, $p = 0.006$), which explains 45% of the variance; for the LB participants ($F(1,14) = 9.12$, $p = 0.010$), it explains 41% of the variance; and for the blindfolded group ($F(1,14) = 29.26$, $p < 0.001$), 69% of the variance. In the sighted group, no association was found ($F(1,14) = 2.14$, $p = 0.167$). Following these results, Fisher's R-to-Z transformation test showed that none of these correlations were significantly different from each other: CB and LB ($z = 0.12$, $p = 0.904$), CB and blindfolded ($z = -0.09$, $p = 0.035$), and LB and blindfolded ($z = 1.06$, $p = 0.289$), suggesting that performance in the deductive reasoning argument task does not depend on vision or blindness onset (see Figure 3). In addition, testing the differences between the sighted controls and the other groups yielded non-significant results in the CB ($z = 1.02$, $p = 0.153$) and LB groups ($z = 0.897$, $p = 0.185$), but yielded a significant result in the blindfolded group ($z = 1.95$, $p = 0.025$).

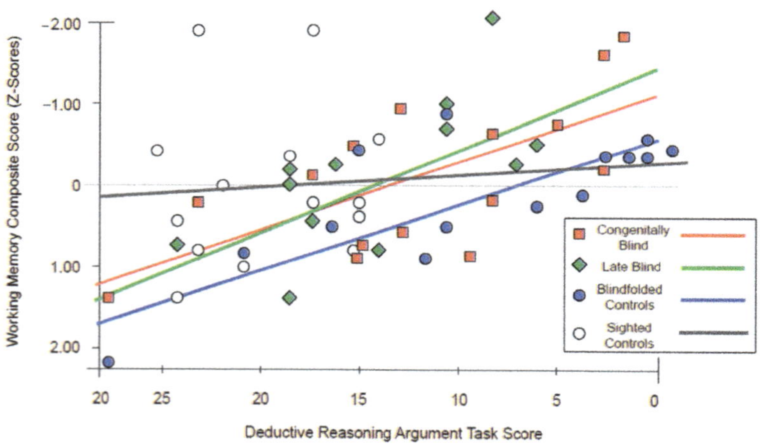

Figure 3. Correlation of reasoning and working memory tasks in the sighted, blindfolded, congenital, and late blindness groups. Scatter plot showing the correlations between deductive reasoning scores (x) and working memory scores (y) for congenitally blind (red squares), late blind (green lozenges), blindfolded controls (blue circles), and sighted controls (white circles). All the correlations between the deductive reasoning argument task and working memory composite score were significant for all the groups, except for the sighted controls.

4. Discussion

Our results show differences in deductive reasoning and working memory in the verbal domain between the sighted, CB, and LB groups. The CB participants outperformed the sighted and LB groups in both of the working memory tasks: the digit span backwards and the letter–number sequencing task. The groups were not different in terms of their deductive reasoning abilities as measured in the word context test. In the deductive reasoning argument task, the blindfolded group performed significantly worse than the LB and sighted participants, while the CB group had lower scores than sighted the participants who were allowed to use pen and paper. Furthermore, we found a link between working memory and deductive reasoning skills in all the groups with blindness, either permanent or temporary and regardless of the level of visual experience. The word context test was not different between the groups. Taken together, our results indicate that visual experience is not necessary for the development of deductive reasoning and working memory skills.

4.1. Reasoning Ability

The variability that we found between the groups in terms of reasoning ability is associated with impairment onset. The LB participants, who lost their vision later in life, are not impaired in terms of their reasoning skills as compared to people with normal vision, whereas the CB group did not differ from the blindfolded group and performed worse than the sighted controls. The sensory compensation hypothesis [12] posits that individuals with blindness can function as well as, or better than, sighted individuals due to improvements in their intact senses. Our study has shown that although they lack the ability to see from birth, they did develop reasoning abilities, which are highly related to vision abilities [28]. However, the performance of those with CB was not consistent, showing that different factors other than mere reasoning ability affected their performance. The nature of the task may account for this inconsistency: While both tasks convey deductive abilities, the deductive reasoning argument task relates to relational reasoning, which is considered to be an aspect of deductive reasoning [44]. This ability refers to the inference of relations between variables from other known relations. The deductive reasoning argument task coincides with these features, therefore also entailing a spatial component, which was shown to be more difficult for persons with blindness [28]. The word context test, on the other hand, does not require such inferences of relations, and is thus easier.

Another finding in regards to reasoning is the poor performance of the blindfolded group in the deductive reasoning argument task compared to other groups. It is easy to understand that a group of people who are used to functioning using vision would perform poorly in a situation when vision is absent. It does not seem that the sighted blindfolded group has lower deductive reasoning abilities, however, because they performed equally well as the other groups in the word context test. Rather, working memory load in the deductive reasoning argument task was higher, therefore impairing their ability to perform well. The fact that the sighted controls who used a pen and paper for the task performed better or equally as well as the other groups indicates that it is not reasoning ability but rather working memory load that affects the level of performance. Indeed, the fact that we found that working memory explains 69% of the reasoning variance in the blindfolded group strengthens this conclusion.

4.2. Working Memory

The working memory scores of the blindfolded group were not significantly worse than those of the other groups, suggesting that this skill could rely on visual experience to initially develop, but once developed, can also function without access to vision. Our findings further show that the CB group outperformed the LB and sighted participants in those tasks. Many other studies have shown superior working memory abilities in individuals with CB [19,23,31,44] compared to sighted people, and in those with LB compared to those with normal vision [18,45].

Although differences between the sighted groups and the blind groups have been previously reported [46], differences in working memory between the CB and LB groups have not been found. It is possible that the taxing demands of everyday functioning without vision from an early age may help develop compensatory cognitive ability mechanisms for working memory resulting in the better performance of those with CB over the LB population [31,46,47].

Due to the inability to acquire visual information, the use of working memory in everyday life is more frequent among the visually impaired compared to the sighted participants. As a result, the performance of the visually impaired is equal to or better than that of the sighted [18,31]. As such, it is possible to infer that the extended use of working memory, together with the duration of the impairment, affects working memory ability.

4.3. Relationship between Visual Experience/Working Memory and Deductive Reasoning Skills

While those with LB did not differ from the sighted participants in the reasoning tasks and also performed worse than those with CB in the working memory tasks, there is still a connection to cognitive skills acquired prior to their impairment [35,40]. Indeed, individuals with late blindness extensively use visual representations for both reasoning [28] and working memory [18,31] in order to maximize their functioning with the sensory information available. These diversified aspects indicate possible qualitative as well as quantitative differences between the CB and LB populations. Consequently, both aspects should be taken into consideration when clinically assessing working memory in the visually impaired population.

The current study further shows that the combined working memory scores of the LB, CB, and blindfolded groups are positively correlated with the deductive reasoning argument task with no difference between the correlations. The sighted participants, however, do not show such an association. This implies that, although there might be differences in working memory and reasoning abilities between visually impaired and sighted people, a moderate extent of the variance in terms of reasoning scores is explained by working memory in the verbal domain [7–9,11]. The fact that the sighted participants who used a pen and paper did not show any correlation between the two abilities indicates that working memory load was reduced significantly, and thus deepens our understanding of its role in reasoning ability. This suggestion should be taken with caution, however, because these differences exist only between the sighted controls and blindfolded groups and not between the sighted controls and both CB and LB groups.

Our results are therefore in line with those of others who have suggested that in order to have good reasoning abilities, one must be able to use working memory adequately, which includes both retaining and manipulating data [7–11]. Therefore, the claim of working memory's involvement in reasoning ability is supported by research in blind and sighted people, also suggesting that the relation between these abilities is not associated with visual experience. Our results further lend support to previous evidence that there are cognitive differences between different etiologies of visual impairment, such that visual experience influences abilities as working memory and reasoning. However, it does not affect their relation, meaning that working memory plays a part in reasoning no matter the level of visual experience. Taken together, the current study does not support the perceptual deficit hypothesis [13], as both clinical groups performed equally as well or better than the sighted controls. Of exception is the deductive reasoning argument task since the experimental conditions differed. On the other hand, our findings do not fully support the sensory compensation hypothesis either [12] since we did not detect superior performance in all the tests or differences between blindness etiologies. Our results seem to support the sensory compensation and brain plasticity hypothesis, which depends on different factors such as etiology of the blindness and the type of abilities measured.

4.4. Future Considerations

Considering the importance of providing accommodations for people with visual impairment [47], other cognitive abilities that were beyond the scope of this study, such as inhibition or cognitive flexibility, may also affect reasoning [8] and further explain the performance of those with LB and CB. Future studies designed to specifically explore other executive functions and their impact on reasoning amongst the visually impaired should make use of a wider battery of tests to achieve a better understanding of the abilities implicated in reasoning. In line with this observation, other populations with visual impairments (e.g., short sightedness and those with damaged visual fields) may benefit from similar studies.

5. Conclusions

The findings of this study demonstrate that reasoning ability in the verbal domain amongst individuals with visual impairment is not uniform, as different tasks yielded different results. Consequently, it is understood that reasoning is highly associated with several dynamic factors: visual impairment onset, level of reasoning difficulty, and degree of working memory load. Each of these factors has an important impact on the performance of visually impaired individuals' reasoning function. Therefore, these findings should be taken into consideration when interpreting the reasoning ability of the visually impaired. Assessments carried out in educational and vocational settings should highlight our findings during evaluation so that adjustments and accommodations for visually impaired individuals can be made accordingly.

Author Contributions: Conceptualization, E.H., N.E. and D.-R.C.; methodology, E.H. and N.E.; formal analysis, E.H. and N.E; writing—original draft preparation, E.H. and D.-R.C.; writing—review and editing, E.H., D.-R.C. and M.P. All authors have read and agreed to the published version of the manuscript.

Funding: This work was supported by the Ariel University Research Authority Absorption Grant #RA1700000192 (DRC), and by a grant from the Canadian Institute of Health Research (MP).

Institutional Review Board Statement: The study was conducted in accordance with the Declaration of Helsinki and approved by the Ethics Committee of Ariel University (904-SOC, 5/9/2016).

Informed Consent Statement: Informed consent was obtained from all subjects involved in the study.

Data Availability Statement: Data are available at: eyalheled@gmail.com.

Acknowledgments: Special thanks to Miguel Chagnon for help with statistical analysis, Department of Mathematics and Statistics, Université de Montréal, Montreal, QC, Canada.

Conflicts of Interest: The authors declare no conflict of interest. The funders had no role in the design of the study; in the collection, analyses, or interpretation of data; in the writing of the manuscript, or in the decision to publish the results.

References

1. Diamond, A. Executive Functions. *Annu. Rev. Clin. Psychol.* **2013**, *64*, 135–168. [CrossRef]
2. Friedman, N.P.; Miyake, A. Unity and diversity of executive functions: Individual differences as a window on cognitive structure. *Cortex* **2017**, *86*, 186–204. [CrossRef]
3. Miyake, A.; Friedman, N.P.; Emerson, M.J.; Witzki, A.H.; Howerter, A.; Wager, T.D. The Unity and Diversity of Executive Functions and Their Contributions to Complex "Frontal Lobe" Tasks: A Latent Variable Analysis. *Cogn. Psychol.* **2000**, *41*, 49–100. [CrossRef]
4. Bacon, A.M.; Handley, S.J.; Dennis, I.; Newstead, S.E. Reasoning strategies: The role of working memory and verbal-spatial ability. *Eur. J. Cogn. Psychol.* **2008**, *20*, 1065–1086. [CrossRef]
5. Ahmed, F.S.; Miller, L.S. Executive Function Mechanisms of Theory of Mind. *J. Autism Dev. Disord.* **2011**, *41*, 667–678. [CrossRef]
6. Chuderski, A.; Necka, E. The contribution of working memory to fluid reasoning: Capacity, control, or both? *J. Exp. Psychol. Learn. Mem. Cogn.* **2012**, *38*, 1689–1710. [CrossRef]
7. Ariës, R.J.; Groot, W.; Van Den Brink, H.M. Improving reasoning skills in secondary history education by working memory training. *Br. Educ. Res. J.* **2015**, *41*, 210–228. [CrossRef]
8. Diamond, A. Understanding Executive Functions: What Helps or Hinders Them and How Executive Functions and Language Development Mutually Support One Another. *Perspect. Lang. Lit.* **2014**, *40*, 7–11.
9. Kyllonen, P.C.; Christal, R.E. Reasoning ability is (little more than) working-memory capacity?! *Intelligence* **1990**, *14*, 389–433. [CrossRef]
10. Olesen, P.J.; Westerberg, H.; Klingberg, T. Increased prefrontal and parietal activity after training of working memory. *Nat. Neurosci.* **2004**, *7*, 75–79. [CrossRef]
11. Simms, N.K.; Frausel, R.R.; Richland, L.E. Working memory predicts children's analogical reasoning. *J. Exp. Child Psychol.* **2018**, *166*, 160–177. [CrossRef] [PubMed]
12. Fine, I.; Park, J.M. Blindness and human brain plasticity. *Annu. Rev. Vis. Sci.* **2018**, *4*, 337–356. [CrossRef] [PubMed]
13. Bell, L.; Wagels, L.; Neuschaefer-Rube, C.; Fels, J.; Gur, R.E.; Konrad, K. The Cross-Modal Effects of Sensory Deprivation on Spatial and Temporal Processes in Vision and Audition: A Systematic Review on Behavioral and Neuroimaging Research since 2000. *Neural Plast.* **2019**, *2019*, 9603469. [CrossRef] [PubMed]
14. Merabet, L.B.; Pascual-Leone, A. Neural reorganization following sensory loss: The opportunity of change. *Nat. Rev. Neurosci.* **2010**, *11*, 44–52. [CrossRef]
15. Chebat, D.R.; Heimler, B.; Hofsetter, S.; Amedi, A. The implications of brain plasticity and task selectivity for visual rehabilitation of blind and visually impaired individuals. In *The Neuroimaging of Brain Diseases*; Springer: Cham, Switzerland, 2018; pp. 295–321.
16. Dormal, V.; Crollen, V.; Baumans, C.; Lepore, F.; Collignon, O. Early but not late blindness leads to enhanced arithmetic and working memory abilities. *Cortex* **2016**, *83*, 212–221. [CrossRef]

17. Heled, E.; Oshri, O. Validation of the Tactual Span in individuals with congenital and acquired blindness. *Br. J. Vis. Impair.* **2021**, 1–15. [CrossRef]
18. Pigeon, C.; Marin-Lamellet, C. Evaluation of the attentional capacities and working memory of early and late blind persons. *Acta Psychol.* **2015**, *155*, 1–7. [CrossRef]
19. Withagen, A.; Kappers, A.M.L.; Vervloed, M.P.J.; Knoors, H.; Verhoeven, L. Short term memory and working memory in blind versus sighted children. *Res. Dev. Disabil.* **2013**, *34*, 2161–2172. [CrossRef]
20. Park, H.J.; Chun, J.W.; Park, B.; Park, H.; Kim, J.I.; Lee, J.D.; Kim, J.J. Activation of the occipital cortex and deactivation of the default mode network during working memory in the early blind. *J. Int. Neuropsychol. Soc.* **2011**, *17*, 407–422. [CrossRef]
21. Rokem, A.; Ahissar, M. Interactions of cognitive and auditory abilities in congenitally blind individuals. *Neuropsychologia* **2009**, *47*, 843–848. [CrossRef]
22. Vecchi, T.; Tinti, C.; Cornoldi, C. Spatial memory and integration processes in congenital blindness. *Neuroreport* **2004**, *15*, 2787–2790.
23. Cohen, H.; Voss, P.; Lepore, F.; Scherzer, P. The Nature of Working Memory for Braille. *PLoS ONE* **2010**, *5*, e10833. [CrossRef] [PubMed]
24. Occelli, V.; Lacey, S.; Stephens, C.; Merabet, L.B.; Sathian, K. Enhanced verbal abilities in the congenitally blind. *Exp. Brain Res.* **2017**, *235*, 1709–1718. [CrossRef] [PubMed]
25. Aleman, A.; van Lee, L.; Mantione, M.H.; Verkoijen, I.G.; de Haan, E.H.F. Visual imagery without visual experience: Evidence from congenitally totally blind people. *Neuroreport* **2001**, *12*, 2601–2604. [CrossRef]
26. Cornoldi, C.; Cortesi, A.; Preti, D. Individual differences in the capacity limitations of visuospatial short-term memory: Research on sighted and totally congenitally blind people. *Mem. Cognit.* **1991**, *19*, 459–468. [CrossRef]
27. Vecchi, T. Visuo-spatial Imagery in Congenitally Totally Blind People. *Memory* **1998**, *6*, 91–102. [CrossRef]
28. Knauff, M.; May, E. Mental imagery, reasoning, and blindness. *Q. J. Exp. Psychol.* **2006**, *59*, 161–177. [CrossRef]
29. Edmonds, C.J.; Pring, L. Generating inferences from written and spoken language: A comparison of children with visual impairment and children with sight. *Br. J. Dev. Psychol.* **2006**, *24*, 337–351. [CrossRef]
30. Knauff, M.; May, E. Visual Imagery in Deductive Reasoning: Results from experiments with sighted, blindfolded, and congenitally totally blind persons. In Proceedings of the Twenty-sixth Annual Conference of the Cognitive Science Society; Lawrence Erlbaum: Mahwah, NJ, USA, 2004; pp. 708–713.
31. Bliss, I.; Kujala, T.; Hämäläinen, H. Comparison of blind and sighted participants' performance in a letter recognition working memory task. *Cogn. Brain Res.* **2004**, *18*, 273–277. [CrossRef]
32. Bonino, D.; Ricciardi, E.; Sani, L.; Gentili, C.; Vanello, N.; Guazzelli, M.; Vecchi, T.; Pietrini, P. Tactile spatial working memory activates the dorsal extrastriate cortical pathway in congenitally blind individuals. *Arch. Ital. Biol.* **2008**, *146*, 133–146.
33. Bottini, R.; Mattioni, S.; Collignon, O. Early blindness alters the spatial organization of verbal working memory. *Cortex* **2016**, *83*, 271–279. [CrossRef]
34. Raz, N.; Striem, E.; Pundak, G.; Orlov, T. Report Superior Serial Memory in the Blind: A Case of Cognitive Compensatory Adjustment. *Curr. Biol.* **2007**, *17*, 1129–1133. [CrossRef] [PubMed]
35. Ruggiero, G.; Iachini, T. The Role of Vision in the Corsi Block-Tapping Task: Evidence From Blind and Sighted People. *Neuropsychology* **2010**, *24*, 674–679. [CrossRef]
36. Cattaneo, Z.; Vecchi, T. Supramodality Effects in Visual and Haptic Spatial Processes. *J. Exp. Psychol. Learn. Mem. Cogn.* **2008**, *34*, 631–642. [CrossRef]
37. Rindermann, H.; Ackermann, A.L.; te Nijenhuis, J. Does Blindness Boost Working Memory? A Natural Experiment and Cross-Cultural Study. *Front. Psychol.* **2020**, *11*, 1571. [CrossRef]
38. Carpenter, P.A.; Eisenberg, P. Mental rotation and the frame of reference in blind and sighted individuals. *Perception, Psychophys.* **1978**, *23*, 117–124. [CrossRef]
39. MacCluskie, K.C.; Tunick, R.H.; Dial, J.G.; Paul, D.S. The role of vision in the development of abstraction ability. *J. Vis. Impair. Blind.* **1998**, *92*, 189–199. [CrossRef]
40. Thinus-Blanc, C.; Gaunet, F. Representation of space in blind persons: Vision as a spatial sense? *Psychol. Bull.* **1997**, *121*, 20–42. [CrossRef]
41. Delis, D.C.; Kaplan, E.; Kramer, J. *Delis-Kaplan Executive Function System: Examiners Manual*; The Psychological Corporation: San Antonio, TX, USA, 2001.
42. Wechsler, D. *Wechsler Adult Intelligence Scale*, 3rd ed.; Psychological Corporation: San Antonio, TX, USA, 1997.
43. Snyder, H.R.; Miyake, A.; Hankin, B.L. Advancing understanding of executive function impairments and psychopathology: Bridging the gap between clinical and cognitive approaches. *Front. Psychol.* **2015**, *6*, 328. [CrossRef]
44. Knauff, M.; Mulack, T.; Kassubek, J.; Salih, H.R.; Greenlee, M.W. Spatial imagery in deductive reasoning: A functional MRI study. *Cogn. Brain Res.* **2002**, *13*, 203–212. [CrossRef]
45. Salillas, E.; Graná, A.; El-Yagoubi, R.; Semenza, C. Numbers in the blind's eye. *PLoS ONE* **2009**, *4*, 3–8. [CrossRef]
46. Wan, C.Y.; Wood, A.G.; Reutens, D.C.; Wilson, S.J. Early but not late-blindness leads to enhanced auditory perception. *Neuropsychologia* **2010**, *48*, 344–348. [CrossRef]
47. Bruce, S.M.; Luckner, J.L.; Ferrell, K.A. Assessment of Students With Sensory Disabilities: Evidence-Based Practices. *Assess. Eff. Interv.* **2018**, *43*, 79–89. [CrossRef]

sensors

MDPI

Article

Network QoS Impact on Spatial Perception through Sensory Substitution in Navigation Systems for Blind and Visually Impaired People

Santiago Real * and Alvaro Araujo

B105 Electronic Systems Lab, Escuela Técnica Superior de Ingenieros de Telecomunicación, Universidad Politécnica de Madrid, Avenida Complutense 30, 28040 Madrid, Spain
* Correspondence: sreal@b105.upm.es; Tel.: +34-91-0672-244

Abstract: A navigation system for individuals suffering from blindness or visual impairment provides information useful to reach a destination. Although there are different approaches, traditional designs are evolving into distributed systems with low-cost, front-end devices. These devices act as a medium between the user and the environment, encoding the information gathered on the surroundings according to theories on human perceptual and cognitive processes. Ultimately, they are rooted in sensorimotor coupling. The present work searches for temporal constraints due to such human–machine interfaces, which in turn constitute a key design factor for networked solutions. To that end, three tests were conveyed to a group of 25 participants under different delay conditions between motor actions and triggered stimuli. The results show a trade-off between spatial information acquisition and delay degradation, and a learning curve even under impaired sensorimotor coupling.

Keywords: QoS; sensory substitution; spatial perception; spatial cognition; navigation; virtual reality; orientation and mobility; wearable

1. Introduction

The penetration of pervasive computing in society has opened new possibilities as to how we interact with the world. For instance, smartphones and low-cost wearables serve as a window to an ever-increasing amount of information, backed by a network that manages the acquisition, storage, and processing of data.

This new paradigm is promoting the development of devices that assist individuals with perceptual and cognitive deficiencies in everyday tasks [1]. These devices gather real-time data of the environment and provide the user with key information to alleviate disabilities, e.g., assisting social interactions in cases of Alzheimer's by providing information tips or guiding visually impaired individuals throughout a route. Ongoing research on this topic focuses on collecting user needs under specific scenarios and translating these to device design.

Although the specific technical requirements vary according to the deficiency and the type of assistance provided, system delay was revealed as a key factor in tasks that required human–environment interactions. Additionally, commercially available devices usually need to be portable and lightweight while withstanding hours of uninterrupted operation under relatively high computational load. Therefore, common designs rely on front-end wireless sensors and low-latency computation offloading [2].

This perspective particularly reflects navigation systems for blind and visually impaired (BVI) individuals [3]. The purpose of such systems is to promote self-sufficiency when finding a destination point in a city, campus, etc., through orientation and mobility assistance.

The most common navigation systems verbally guide the user throughout a route, relying on GNSS or a beacon infrastructure to locate the user outdoors and indoors, as seen in Lazzus, BlindSquare and NavCog smartphone apps. This is usually enriched with data of nearby points of interest recorded in remote sources, such as Google Places or OpenStreetMap. Conversely, other systems provide assistance in close-range scenarios, allowing the user to identify and react to potentially hazardous elements in their paths. The usage of sensors varies, but it usually includes cameras and depth sensors to perform simultaneous locating and mapping (SLAM) [4] or object identification, e.g., applying pre-trained convolutional neural networks such as YOLO [5,6], while benefiting from edge-computing schemata [4,6].

Even within the field of perceptual and cognitive assistance, system delay constitutes a critical technical requirement for navigation purposes. For instance, system responses in the order of tenths of a second might be needed so that the user can avoid moving elements in their path. However, there is another aspect that might impose even stricter timing constraints: the human–machine interface.

As the amount of available data grows, the need for effective and efficient media to provide spatial information to the user has become more apparent. The study of low-level, nonvisual human perceptual and cognitive processes has revealed another foundational element of navigation systems. In this context, the pioneering work of Bach-y-Rita et al. introduced a new approach: the images captured by a camera could be encoded as tactile stimuli in a test subject's skin, allowing for the recognition of remote elements. This assimilation of information related to a specific sensor modality through another was coined as "sensory substitution," although later studies argued against the suitability of the term [7].

For that distal attribution to happen, the users needed to manage the camera themselves to identify the "contingencies between motor activity and the resulting changes in tactile stimulation" [8]. In line with this, recent theories view human cognition as deeply grounded in action [9]. Particularly, the sensorimotor contingencies (SMCs) theory revised sensorimotor coupling as a bidirectional process—in opposition to classic unidirectional models—that is constitutive of perception.

Overall, in navigation assistance for BVI individuals, new spaces of human–world interaction are created that ultimately rely on sensorimotor coupling. Due to their technical implementation, the navigation systems are subjected to multiple sources of delay, e.g., propagation speed, spectrum saturation, traffic congestion, computing time, etc., most of which are aggregated by the end-to-end communication quality of service (QoS).

In line with this, the degradation of the QoS, emphasizing communication latency and jitter, is expected to have a negative impact on such human–machine interfaces, and in turn system performance.

This design approach could be extended to future perceptual and cognitive assistance design. In that regard, several methodologies grouped under the term "quality of experience (QoE) [10] have been developed to assess proper operation of the system from the perspective of the user.

For instance, the ITU's recommendations when reproducing 360° video in head-mounted displays [11] include asking the test subject about sickness symptoms or even evaluating exploratory movements with eye tracking. Additionally, performance-based tests are suggested for task-oriented applications, e.g., recognition of individuals from a video stream for surveillance purposes [12]. In general, subjective QoE analysis allows the development of system-performance prediction models from user behavior and objective QoS measurements (e.g., [13,14]). However, no prior work was found in relation to immersive human–machine interfaces or navigation assistance as mechanisms to counter visual disabilities.

In this context, the present contribution focuses on analyzing potential constraints of system delay in navigation assistance that derive from the human–machine interface.

To that end, a mixed-reality tool was developed to perform simple perceptual and cognitive experiments with sensory substitution devices (SSDs) under different system-delay scenarios. The experiments were designed taking into consideration the main purpose of a navigation system, i.e., assisting in orientation and mobility tasks throughout a route. In line with this, the measured variables are related to key user navigation performance indicators.

2. Materials and Methods

To analyze the impact of system delay on navigation systems, two SSDs were tested in immersive virtual environments. The system delay was degraded according to previously modeled stochastic processes. Finally, different user-performance measurements were recorded.

In the following subsections, updated versions of the navigation systems' test-bench tools Virtually Enhanced Senses (VES) [15] and VES-PVAS SSD [16] are briefly presented. Thereafter, the three perceptual and cognitive tests conducted are described.

2.1. The Virtually Enhanced Senses System

The VES system is a wireless, mixed-reality platform developed to implement and test complete navigation systems (Figure 1). It immerses the user in virtual or previously scanned real environments through a visual–inertial motion-capture system. Additionally, VES allows the implementation of a wide range of sensory substitution devices, degrading network quality of service (QoS) according to previously modeled stochastic processes, and taking user behavioral data regarding navigation performance.

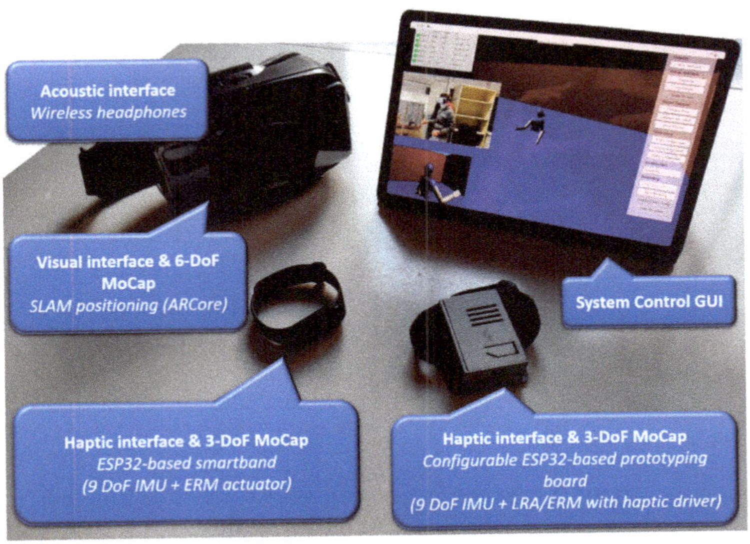

Figure 1. The VES system (the image in the tablet is a capture of the GUI and a photo of the user).

The controlled QoS degradation is a new feature embedded in a custom communication stack over UDP/IP. Following a publisher-subscriber pattern, streaming data such as user motion or motor driving signals is shared among devices as time-stamped packages. Each device queues the incoming stream data and discards or delays each package according to a previous simulation of end-to-end communication. Therefore, it supports simultaneous data streams with different conditions of jitter and packet loss, which in turn results from the simulation of several protocol stacks under specific conditions of traffic load, spectrum occupation, etc.

Nevertheless, this solution has two drawbacks. Firstly, it operates over existing wireless communications, and therefore there is a QoS baseline. Secondly, the implementation of VES over a non-real-time OS such as Windows might include unavoidable jitter. However, this approach is considered enough for testing purposes if the time resolution is maintained over 20–30 ms.

2.2. VES-PVAS Sensory Substitution

VES-PVAS is a virtual SSD based on the Virtual Acoustic Space (VAS) project. Following its predecessor, this SSD captures the 3D surfaces within a virtual camera's field of view (FoV) and sequentially reproduces short-duration spatial virtual sound sources over it, i.e., stereo pixels (Figure 2). These stereo pixels take advantage of natural hearing perceptions of distance or material composition or even the size of an element through the number of sound sources triggered.

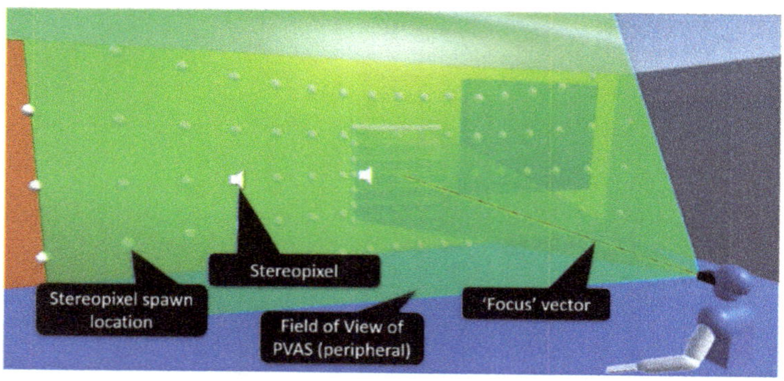

Figure 2. VES-PVAS sensory substitution system.

This kind of SSD design approach is advantageous in terms of both processing load and benchmarking, as all user–environment interactions are fully characterized by a limited set of moving spatial points that trigger stimuli—e.g., stereo pixels—and the user's motion. In cases of mixed or virtual reality approaches, the graphical rendering is not strictly required. The density of points can be adjusted dynamically according to the required spatial resolution.

The previous VES-PVAS allowed configuration of the FoV, the N × M matrix of spawning points for the stereo pixels, the detection range, and the seed for the pseudo-aleatory spawn sequence criteria of the stereo pixels within the FoV. In addition to this, the current version includes the following.

- Configurable segmentation. All elements in the virtual environment can be freely grouped and mapped to specific configuration and spawn rules for the stereo pixels. This feature builds on previous attempts to embed Gestalt-like laws of visual perception in SSDs [8,17].
- Figure–background discrimination. Building on the configurable segmentation feature, the elements within the FoV are identified as "focused" or "not focused." Specifically, only the element that occupies the center of the FoV (Figure 2) is considered "focused." Thereafter, all the stereo pixels triggered by focused and unfocused elements are added a constant volume gain accordingly.
- Peripheral and foveal vision. Two overlaying VES-PVAS SSDs were configured: the first one uses a relatively wide FoV and low spatial resolution, whereas the second one is set with a narrow FoV and high spatial resolution, as shown in Figure 2.

In early tests, the elevation of the stereo pixels seemed difficult to locate through the head-related transfer function (HRTF) module. Therefore, taking advantage of an altitude-to-pitch cross-modal correspondence, the stereo pixels' sound pitch was modulated according to its relative altitude within the FoV. This is modeled by the following expression:

$$A_{out}(f) = A_{in}\left(f * \left[1 + \frac{1}{2}\tan\alpha\right]\right)$$

where *A*(*f*) is the audio signal spectrum and α the relative altitude within the FoV in degrees.

2.3. Experimental Procedures

As described in previous sections, the objective is to assess the impact of system delay in navigation systems for BVI individuals. Overall, the aim is to obtain objective measurements related to navigation performance and analyze how these measurements vary under different scenarios of system delay. In this regard, the first challenge consists in choosing appropriate user-testing methods [18].

Starting from the premise of active perception, test subjects must be allowed some degree of freedom when interacting with the environment. This would be key for the user to develop sensorimotor strategies [19], which are extended to locomotion if the spatial features of the environment cannot be apprehended from a single point. However, complete navigation experiments that require moving through an unknown environment with an artificial SMC apparatus, in this case the SSD, are sensitive to multiple variables that would outweigh the role of the system delay.

Given this context, the methods described herein were conceived as unitary tests. Simple perceptual and cognitive processes useful for navigation are evaluated in virtual environments with only 2–3 elements that can be perceived from a single vantage point. Three tests were conducted: relative width estimation, symbol discrimination, and finally, search and focus a single moving element.

The focus of the present contribution is placed on the temporal restrictions on system delay deriving from the sensorimotor coupling. To that end, an artificial delay was introduced in all three tests between the user's motion and the resulting haptic/acoustic stimuli. Early tests revealed that approximately 500 ms produced noticeable effects in the measured performance indicators. In line with this, three models of system delay were tested: no delay, constant delay of 500 ms, and a variable delay following a Gaussian distribution of N (500, 100) ms.

No prior SSD training was done. Instead, the experiments were designed with increasing difficulty (Table 1), taking advantage of the relatively low cognitive load and intuitiveness of the SSD under test. This approach has a double purpose: firstly, it serves to homogenize the initial conditions for all users, and secondly it allows observation of the learning curve and any hypothetical adaptation to sensorimotor coupling impairment as the system delay degraded.

Table 1. Order of the experiments.

Order of The Tests		Sensorimotor Delay		
		None	500 ms	N (500, 100) ms
Relative width estimation	Ratio 1.70	1	4	7
	Ratio 1.45	2	5	8
	Ratio 1.30	3	6	9
Symbol discrimination		10	11	12
Search and focus on moving element		13	14	15

The tests were conducted on a heterogeneous sample of 22 normal-sighted and 3 BVI users aged 19–72 years, with a male-female ratio of 17-8. In the BVI group, the first test subject lacked peripheral vision, while the remaining subjects were completely blind.

Although the sample size was not sufficient to get any statistical significance, it served as a first step in the study of the users' behavioral patterns.

The instructions for the tests were provided to the normal-sighted users in a video tutorial. BVI users were given a verbal explanation.

Further details on the three tests are included in the following subsections.

2.3.1. Relative Width Measurement

This test addresses the field of relative distance estimation. It starts from the premise that this is a basic element in the development of mental representations of space, and it holds a key role in both navigation and mobility tasks. Nevertheless, recent research questions the extent of spatial knowledge that is required for navigation [20].

The SSD used, from now on referred as "virtual cane," consists of a virtual proximity sensor with a relatively narrow FoV (~10°) that triggers haptic stimuli in the user's hand once an element enters its detection range (Figure 3).

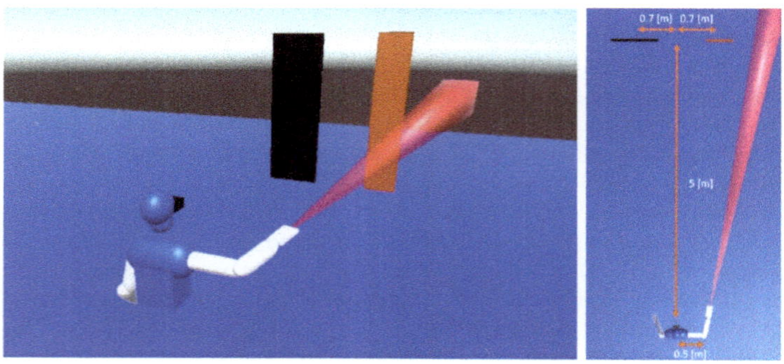

Figure 3. Test of relative width with a haptic SSD. The detection area of the SSD is in red.

This type of SSD has been used extensively since the first developed electronic travel aids, including a variety of commercial products. This extends to currently available devices and ongoing research on BVI navigation. Taking advantage of its relatively low cognitive load, the information provided is usually enriched with auditory feedback [21] or by relying on multiple sensors distributed over the user's body [22].

Although the virtual cane encodes distance measurements as vibration intensity, in the following experiments, the distance variations are considered negligible; therefore, the SSD will only activate or deactivate the haptic actuator (ERM motor) according to the presence or absence of elements at the pointed direction. Additionally, in the current configuration of the motor driver, the start and brake times are measured under 100 ms, adding to the system delay.

As for the methods to evaluate distance estimates, a simple paired comparison was preferred, as no external metric was required. The distance measurements would need few sequences of movements, as well as low usage of working memory and inferential processing. Thus, the system delay was expected to play a major role in the obtained data, which were reduced to hit rates that could be analyzed from an above-chance perspective.

Once the test begins, the user is immersed in a virtual scenario with two rectangles 5 m in front (Figure 3). As for their dimensions, three variations were presented to study a possible trade-off between sensorimotor delay and spatial resolution when developing a mental representation of the environment. With a base dimension of 0.7×3.5 m, three relative widths were tested: 1.3, 1.45 and 1.7, which corresponds to 0.9, 1 and 1.3 m respectively–. For each comparison, the rectangles swapped positions at random. Finally, the time to answer was also recorded to observe possible variations under different test conditions.

2.3.2. Symbol Discrimination

The user is asked to use the virtual cane to identify a symbol from a known set (Table 2), provided as 3D-printed embossed images. The purpose is to check whether specific spatial features from the environment can be recognized, e.g., building on distance estimations made with an SSD. The virtual scenario is the same as in the previous test (Figure 3).

Table 2. Set of symbols presented to the user.

Symbol	A	B	C	D	E	F
Figure						
Dimensions [m]	4.5×2.5	4.5×2.5	4.5×2.5	4.5×2.5	4×4	4.5×2.5

Originally, the symbols were designed to check if the user could discriminate the number of elements and their relative dimensions with simple horizontal and vertical movements. This was useful to evaluate potential "proximity" between symbols from the perspective of the user estimations. This in turn relates to the specific features of the sensorimotor apparatus (SSD).

Early tests were conducted to adjust the size and content of the set. Finally, the six symbols included in Table 2 were considered enough for this contribution. The symbol under **E** was included to increase the difficulty, as it forces higher-precision movements for effective discrimination.

Once the test starts, the user is presented with a random permutation of all 6 elements of the set. This is repeated for all sensorimotor delay scenarios. Again, the experiments follow a common order for all users (Table 1), in which the system delay is degraded progressively. The hit rate and time to answer are recorded for all scenarios.

2.3.3. Search and Focus on a Moving Element

Finally, the third test was conceived as a performance analysis of SSD that encodes stream data from a camera. This constitutes one of the main families of SSD, which includes well-known visual-to-auditory SSD projects such as vOICe, EyeMusic [23], or VAS, as well as visual-to-tactile SSD, e.g., BrainPort [24], etc.

It is noted that these systems are especially prone to incur an additional system delay due to bandwidth differences between sensory modalities. This point is exemplified by vOICe and EyeMusic, in which low-resolution images are converted into audio signals with durations in the order of seconds. This acts as a baseline sensorimotor coupling impairment, which can be reduced at the cost of spatial resolution.

Analogously to the visual sense, this family of camera-based SSD provides simultaneous perception of different elements within the FoV. In navigation tasks, one of the main advantages is that the user is informed about the presence and relative position of elements in a relatively large area. In line with this, navigation performance is evaluated with a simple perceptual test: pointing to a single element.

The user is asked to keep track of a moving element that follows a predefined sequence of positions unknown to the user. For this test, the equipment consists of a VR headset with a 6 DoF MoCap module and a couple of headphones. The specific configuration of VES-PVAS is included in the following table (Table 3).

The sequence is composed of 9 positions arranged in a 3×3 matrix, as shown in Figure 4. The element covers all 8 possible trajectories, starting from the center. Once the test starts, the element stay in place until the user maintains their focus steadily for 3 s through VES-PVAS. Thereafter, it moves to the next position within the gray surface, i.e., screen, and so on. This cycle is repeated for all 3 scenarios of sensorimotor delay. Finally, the time to focus (TTF) and the user's gaze trajectories are recorded as objective performance variables.

Table 3. VES-PVAS configuration of the third test.

PVAS Configuration	Peripheral	Foveal
M	17	17
N	7	7
FoV_x	100°	20°
FoV_y	50°	20°
Detection distance	20 m	20 m
Seed	43	43
Period	20 ms	20 ms
"Focused" volume gain	−1 dB	0 dB
"Not focused" volume gain	−3.5 dB	−9 dB

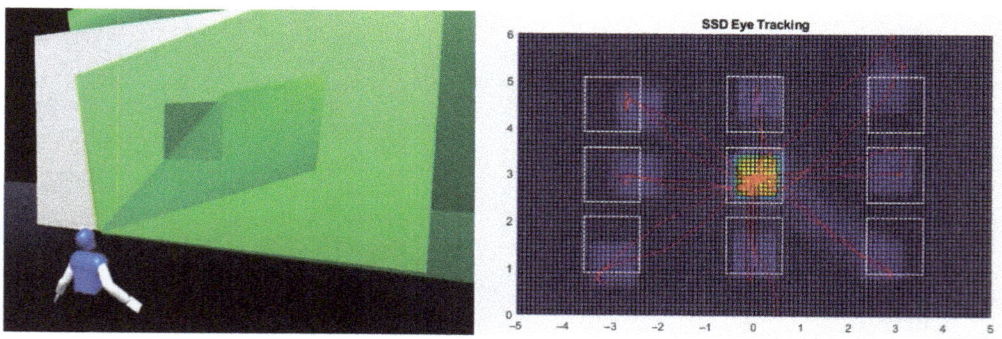

Figure 4. Test of search and focus a moving element. In the right image, mock SSD eye tracking is presented.

The positions of the element were chosen according to the following criteria.

- The element should always be perceivable by the user, i.e., stay within the FoV. This eliminates the need for exploratory movements, which would otherwise constitute a parasite effect included in the measured variables.
- The test must favor intentionality in the user–environment interaction. In line with this, it was concluded that the element motion should include uncertainty in at least two axes. This serves to avoid focusing the element by chance or with simple sweeping movements independently of the SSD feedback.
- The SSD feedback is different when providing information regarding the X and Y relative position. Therefore, various combinations of X–Y trajectories should be included to assess any potential impact on the performance indicators.

The users are expected to unfold the SSD feedback without previous training, as experienced in previous work [11]. Thereafter, the TTF is directly related to performance: it would increase in scenarios with higher difficulty, e.g., under system delay degradation. Conversely, it would decrease with user experience.

On the other hand, the gaze tracking shows the user motion when focusing on an element. It is considered that in the best case, it would follow a straight line from the initial to the ending position. Under system delay, a deviation is expected.

3. Results

In total, 25/25, 23/25, and 23/25 of the users were able to complete the first, second, and third tests, respectively. This accounts for the intuitiveness of the SSD implemented, as well as the embedding of SSD training in the user testing. On the other hand, the results of the BVI and normal-sighted groups showed no significant difference. Therefore, the following data aggregate all test subjects indistinctly.

The hit rate and time to answer for the relative width measurement and symbol discrimination are shown in Figure 5. This figure includes the average results from the first two tests, divided into experiments 1–12, as described in Table 1.

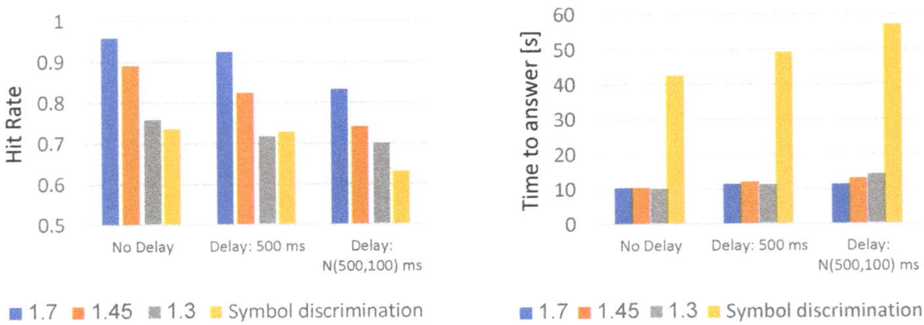

Figure 5. Average hit rate and time to answer in the relative width measurement and symbol identification tests.

For all 12 experiments making up the three tests (Table 1), the results were above chance. In the first test (1–9), the hit rate in all three paired comparisons denoted a general trade-off between spatial resolution and system delay. Conversely, the time to answer showed a significant increment only in symbol discrimination as the delay degraded.

All users reported that the elements in the virtual scenario "moved" as the system delay increased, and some even related that to their own movement, but none seemed to associate it with a delay in the stimuli triggering.

In relation to the second test, the questionnaire answers are gathered in the following confusion matrices (Figure 6). On top of the hit rate, these matrices show the most common symbol-guess errors made by the users. From left to right, each of the matrices corresponds to a different system delay scenario.

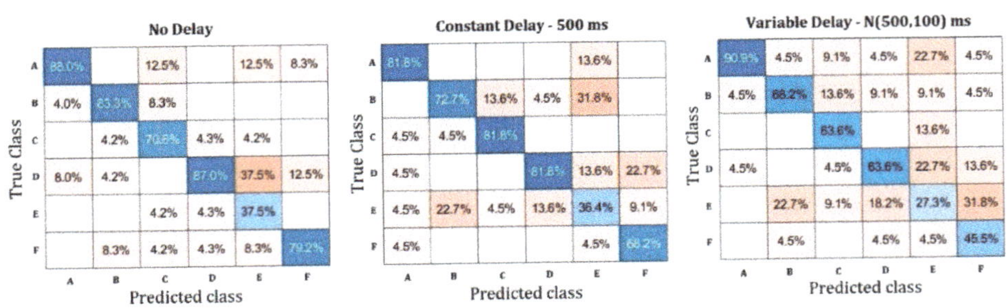

Figure 6. Confusion matrices in the symbol identification test.

As could be anticipated, each symbol showed different hit rates and statistical proximity to the other symbols in terms of user estimations. However, despite showing analogous geometric features, the pairs A–B (mirroring) and C–F (rotation) results differed. On the other hand, the element E exhibited the worst performance results for all three delay scenarios.

Outside the methods specified, sensorimotor strategy development was also observed. When using the virtual cane, almost all users seemed to measure distances through vibration duration as they moved the cane at a constant speed along the X or Y axis. The accuracy of those movements showed key importance throughout the tests. It could even be used to advance the user answers to the questionnaires.

Figure 7 shows the average gaze-tracking heatmap of the user in test 3. This heatmap is a 2D histogram of the users' gaze as it moved within the gray screen shown in Figure 4. This figure included the starting and ending position of the moving element as white and red squares, respectively. Finally, the 24 graphs correspond to all 8×3 combinations of element trajectory and system delay.

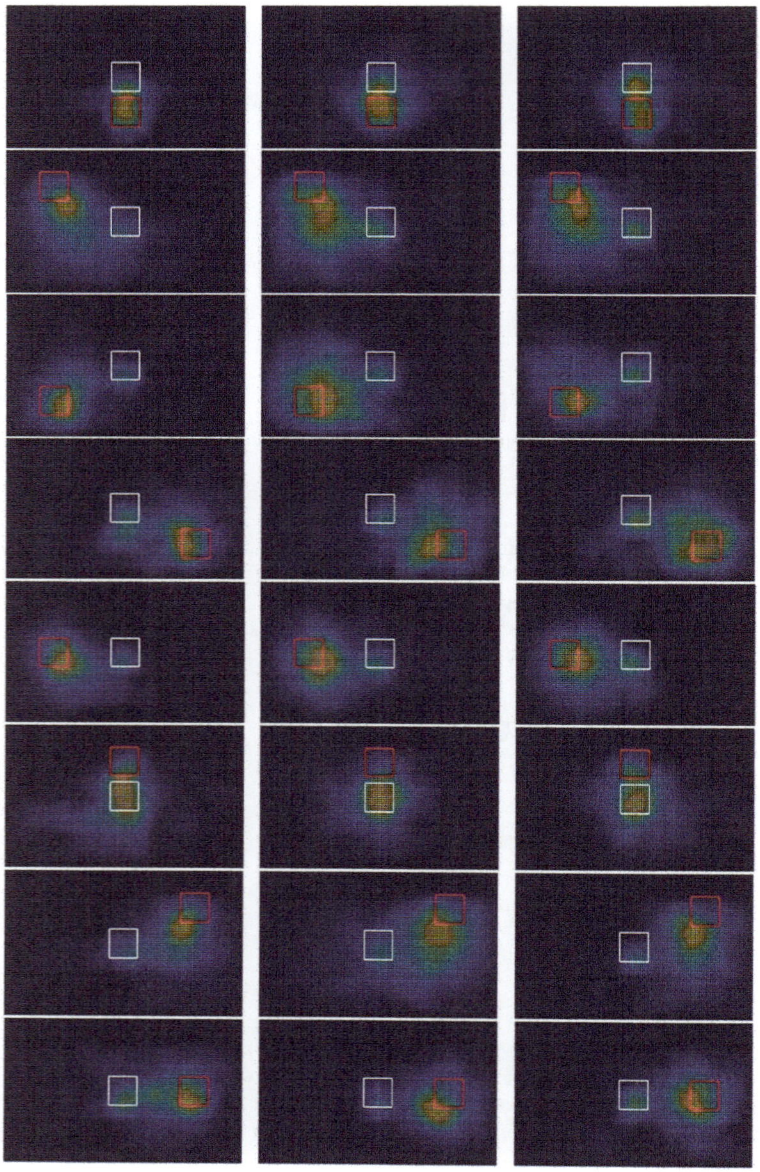

Figure 7. VES-PVAS average foveal gaze-tracking heatmaps of test 3. Each column corresponds to all element trajectories, starting from the center, under the conditions of no delay, 500 ms delay and N (500, 100) ms delay. The white and red boxes represent the starting and ending position of the element.

In all experiments, the "gaze" was always centered close to the target. However, the gaze was relatively dispersed in diagonal trajectories and degraded sensorimotor delay scenarios.

After a few experiments, the users tended to wait for the peripheral stereo pixels to trigger. In line with this, the heatmap shows higher intensity at the center. This result shows an adaptation to the relatively low temporal resolution of camera-based SSD. In particular, the current configuration of VES-PVAS requires approximately 2.4 s to cover all spawn points of the stereo pixels. Nevertheless, this SSD offers a time-resolution trade-off by distributing the positions of the stereo pixels' spawning sequence, based on the traditional interlaced video. This feature serves to accelerate the feedback of relatively large elements within the FoV.

These data are complemented by the TTF, that is, the time required by the test subjects to focus on the moving element. In turn, this time has been divided according to two different events: "first focus" and "maintained focus," i.e., that which triggers the element motion. This can be observed in the following figure.

In Figure 8, the exploratory head movements can be noted: the space between events shows "spikes" in VES-PVAS rotational speed as the user tries to center the element in the foveal region. Once the user focuses on an element, the rotational speed diminishes as he/she tries to maintain it at the center of the FoV.

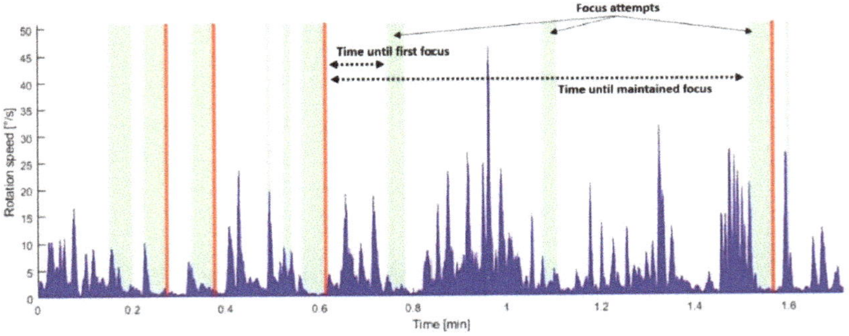

Figure 8. Gaze motion data of a single user in the third test. It includes rotation speed of VES-PVAS FoV (blue): "element focused" (green bars) and "element moved" events (red bars).

The next figure presents the average (Figure 9, left) and standard deviation (Figure 9, right) values of the TTF corresponding to all eight trajectories of the moving element, following the order of the user experiments (Table 1).

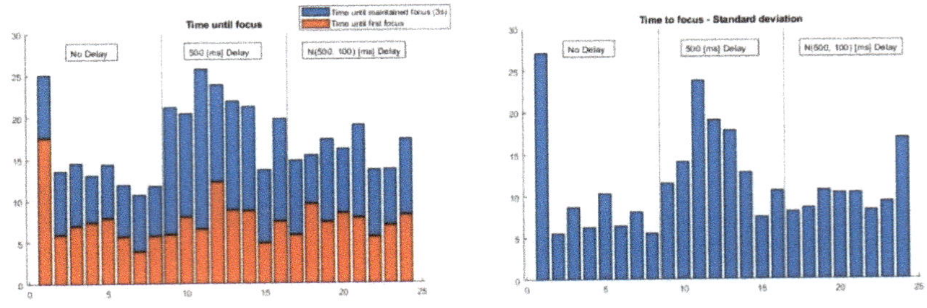

Figure 9. Time inverted until the element was steadily focused (3 s) per trajectory: average (**left**) and standard deviation (**right**) values.

As can be observed, with no prior explanation regarding VES operation, the first position was difficult to detect: it exhibits large average and dispersion values. After that, the TTF converged for all trajectories and users, given the relatively low standard deviation. Once the half-second delay was added, the TTF increased abruptly and with high variance among users (Figure 9, right). Nevertheless, it decreased over time, and stabilized even after the addition of jitter.

Finally, the number of "element-focused" events until the element moved to the next position, i.e., focus attempts (Figure 8), is included in Figure 10. In contrast to the TTF, it increased constantly as the system delay degraded, with no evident signals of user compensation.

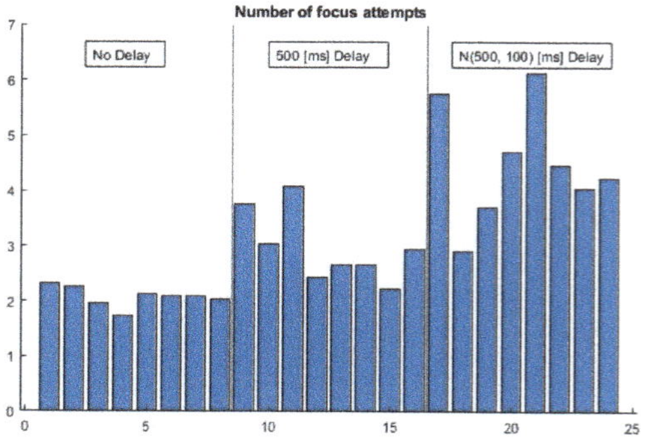

Figure 10. Number of times that the element was focused until it moved to the next position.

4. Discussion

Overall, the under-second system delay added in the sensorimotor coupling was enough to noticeably degrade the modeled navigation performance indicators. This applies to both the SSD and the three conducted tests. Nevertheless, some sort of compensation mechanism was recognized in VES-PVAS.

The first remarkable point is the trade-off between system delay and spatial resolution observed from the hit rates in the first test. The well-known virtual cane approach denoted poor performance in terms of relative distance estimation and identification of primitive figures, which was more apparent as the delay increased and the spatial resolution lowered. Specifically, the latter effect could be observed from the users' poor discrimination rates of figures of relatively little dimension differences. However, it should be noted that the actuator that triggered the tactile stimuli was an ERM motor driven by pulse width modulation PWM signals. Therefore, the ~100 ms maximum time for the start and brake times are noticeable by the user, and act as a low-pass temporal and spatial filter when the user interacts with the environment.

Additionally, the users reported that the element moved, translating the internal effect of an impaired sensorimotor coupled with a distorted representation of the environment. Some of them even accelerated their movements to catch the presumably moving element. This suggests that the temporal requirements for navigation purposes, and specifically for relative position estimation, are even more restrictive.

On the other hand, the experiments endorse the hypothesis that the acquisition of spatial information is exteriorized through the user's interaction with the environment. As described in the previous section, the questionnaire answers could be anticipated after careful examination of the movements of the user within the virtual scenario and the corresponding stimuli generation. This approach to the analysis of sensorimotor strategies could be used to further improve SSD design, propose objective performance parameters, or even adjust the human-machine interface with human-in-the-loop strategies.

As for the third test, one of the most remarkable results is the fast-learning curve in the usage of VES-PVAS. Almost all users were able to focus the elements with no prior training or explanation regarding the sound patterns. Furthermore, the gaze was concentrated in the surroundings of the target for all combinations of element trajectories and sensorimotor delay scenarios.

After adding sensorimotor delay, a second learning curve was seen in which the users adapted their movements according to the new perceptual conditions. This seems to occur in opposition to the results of the virtual cane SSD, in which the internal sensorimotor impairment was translated to external sources and no compensatory behaviors could be observed from the data gathered.

5. Conclusions

The present contribution points out novel timing restrictions of system delay when providing navigation assistance to BVI individuals. These restrictions are derived from human perceptual and cognitive processes, and in turn sensorimotor coupling. Consequently, this is another factor to be taken into consideration in future networked designs in which end-to-end communication delay plays a major and unavoidable role.

The results suggest that there are not hard timing restrictions, but a trade-off between the detail and amount of spatial information that can be provided and the level of degradation of the system delay. This trade-off could even benefit from training, as the test subjects showed a learning curve even under a half-second system delay.

Finally, this contribution proposes a novel approach to human–machine interface design based on objective user-performance parameters gathered from an artificial SMC apparatus. Specifically, it was used to assess the impact of motion-to-photon latency in nonvisual, immersive human–machine interfaces. Overall, this is expected to be useful in the development of future devices for perceptual and cognitive assistance.

Author Contributions: Conceptualization, S.R. and A.A.; methodology, S.R.; software, S.R.; validation, S.R. and A.A.; formal analysis, S.R.; investigation, S.R.; resources, A.A.; data curation, S.R.; writing—original draft preparation, S.R.; writing—review and editing, A.A.; visualization, S.R.; supervision, A.A. All authors have read and agreed to the published version of the manuscript.

Funding: This research received no external funding.

Institutional Review Board Statement: Ethical review and approval were waived for this study due to the non-sensitive nature of the user behavioral data herein presented.

Informed Consent Statement: Test subject consent was waived due to the non-sensitive nature of the data retrieved.

Data Availability Statement: The raw results of the user tests are available online at (DOI: 10.17632/zzswztx4cz.1) and http://elb105.com/ves/ (accessed on 14 March 2023).

Conflicts of Interest: The authors declare no conflict of interest.

References

1. Ha, K.; Chen, Z.; Hu, W.; Richter, W.; Pillai, P.; Satyanarayanan, M. Towards wearable cognitive assistance. In Proceedings of the 12th Annual International Conference on Mobile Systems, Applications, and Services, Bretton Woods, NH, USA, 16–19 June 2014; pp. 68–81.
2. Muñoz, M.O.; Klatzky, R.; Wang, J.; Pillai, P.; Satyanarayanan, M.; Gross, J. Impact of delayed response on wearable cognitive assistance. *PLoS ONE* **2021**, *16*, e0248690.
3. Real, S.; Araujo, A. Navigation Systems for the Blind and Visually Impaired: Past Work, Challenges, and Open Problems. *Sensors* **2019**, *19*, 3404. [CrossRef] [PubMed]
4. Zhang, H.; Ye, C. An Indoor Wayfinding System Based on Geometric Features Aided Graph SLAM for the Visually Impaired. *IEEE Trans. Neural Syst. Rehabil. Eng.* **2017**, *25*, 1592–1604. [CrossRef] [PubMed]
5. Wang, Z.; Li, H.; Chen, J.; Chai, X.; Zhai, Z. A Wearable Vision-To-Audio Sensory Substitution System Based on Deep Learning for the Visually Impaired. In Proceedings of the 2021 International Conference on Digital Society and Intelligent Systems (DSInS), Chengdu, China, 19–21 November 2021; pp. 283–286.
6. Yuan, Z.; Azzino, T.; Hao, Y.; Lyu, Y.; Pei, H.; Boldini, A.; Mezzavilla, M.; Beheshti, M.; Porfiri, M.; Hudson, T.E.; et al. Network-Aware 5G Edge Computing for Object Detection: Augmenting Wearables to "See" More, Farther and Faster. *IEEE Access* **2022**, *10*, 29612–29632. [CrossRef]

7. Lenay, C.; Gapenne, O.; Hanneton, S.; Marque, C.; Genouell, C. Sensory substitution: Limits and perspectives. *Touching Knowing* **2003**, 275–292. [CrossRef]
8. Loomis, J.M.; Klatzky, R.L.; Giudice, N.A. Sensory substitution of vision: Importance of perceptual and cognitive processing. In *Assistive Technology for Blindness and low Vision*; CRC Press: Boca Raton, FL, USA, 2012; pp. 162–191.
9. Engel, A.K.; Maye, A.; Kurthen, M.; König, P. Where's the action? The pragmatic turn in cognitive science. *Trends Cogn. Sci.* **2013**, *17*, 202–209. [CrossRef] [PubMed]
10. ITU. Recommendation P.10: Vocabulary for Performance and Quality of Service and Quality of Experience. Telecommun. Stand. Sect. ITU-T 2017, 100, 1–42. Available online: https://www.itu.int/rec/T-REC-P.10 (accessed on 14 March 2023).
11. ITU-T. Recommendation P.919: Subjective Test Methodologies for 360° Video on Head-Mounted Display. 2020, 1–38. Available online: https://www.itu.int/rec/T-REC-P.919/en (accessed on 14 March 2023).
12. ITU-T. Recommendation P.912: Subjective Video Quality Assessment Methods for Recognition Tasks. 2016. Available online: https://www.itu.int/rec/T-REC-P.912 (accessed on 14 March 2023).
13. Reichl, P.; Egger, S.; Schatz, R.; D'Alconzo, A. The logarithmic nature of QoE and the role of the Weber-Fechner law in QoE assessment. In Proceedings of the 2010 IEEE International Conference on Communications, Cape Town, South Africa, 23–27 May 2010.
14. Steinbach, E.; Strese, M.; Eid, M.; Liu, X.; Bhardwaj, A.; Liu, Q.; Al-Ja'Afreh, M.; Mahmoodi, T.; Hassen, R.; El Saddik, A.; et al. Haptic codecs for the tactile internet. *Proc. IEEE* **2019**, *107*, 447–470. [CrossRef]
15. Real, S.; Araujo, A. VES: A Mixed-Reality Development Platform of Navigation Systems for Blind and Visually Impaired. *Sensors* **2021**, *21*, 6275. [CrossRef] [PubMed]
16. Real, S.; Araujo, A. VES: A Mixed-Reality System to Assist Multisensory Spatial Perception and Cognition for Blind and Visually Impaired People. *Appl. Sci.* **2020**, *10*, 523. [CrossRef]
17. Cronly-Dillon, J.; Persaud, K.; Gregory, R.P.F. The perception of visual images encoded in musical form: A study in cross-modality information transfer. *Proc. R. Soc. Lond. Ser. B Biol. Sci.* **1999**, *266*, 2427–2433. [CrossRef] [PubMed]
18. Schinazi, V. *Representing Space: The Development, Content and Accuracy of Mental Representations by the Blind and Visually Impaired;* University College London: London, UK, 2008.
19. Bermejo, F.; Di Paolo, E.; Hüg, M.X.; Arias, C. Sensorimotor strategies for recognizing geometrical shapes: A comparative study with different sensory substitution devices. *Front. Psychol.* **2015**, *6*, 679. [CrossRef] [PubMed]
20. Lobo, L.; Nordbeck, P.C.; Raja, V.; Chemero, A.; Riley, M.A.; Jacobs, D.M.; Travieso, D. Route selection and obstacle avoidance with a short-range haptic sensory substitution device. *Int. J. Hum. Comput. Stud.* **2019**, *132*, 25–33. [CrossRef]
21. Maidenbaum, S.; Hanassy, S.; Abboud, S.; Buchs, G.; Chebat, D.-R.; Levy-Tzedek, S.; Amedi, A. The "EyeCane", a new electronic travel aid for the blind: Technology, behavior and swift learning. *Restor. Neurol. Neurosci.* **2014**, *32*, 813–824. [CrossRef] [PubMed]
22. Cassinelli, A.; Sampaio, E.; Joffily, S.B.; Lima, H.R.S.; Gusmao, B.P.G.R. Do blind people move more confidently with the Tactile Radar? *Technol. Disabil.* **2014**, *26*, 161–170. [CrossRef]
23. Abboud, S.; Hanassy, S.; Levy-Tzedek, S.; Maidenbaum, S.; Amedi, A. EyeMusic: Introducing a "visual" colorful experience for the blind using auditory sensory substitution. *Restor. Neurol. Neurosci.* **2014**, *32*, 247–257. [CrossRef] [PubMed]
24. Grant, P.; Spencer, L.; Arnoldussen, A.; Hogle, R.; Nau, A.; Szlyk, J.; Nussdorf, J.; Fletcher, D.C.; Gordon, K.; Seiple, W. The Functional Performance of the BrainPort V100 Device in Persons Who Are Profoundly Blind. *J. Vis. Impair. Blind.* **2016**, *110*, 77–89. [CrossRef]

sensors

Article

The Unfolding Space Glove: A Wearable Spatio-Visual to Haptic Sensory Substitution Device for Blind People

Jakob Kilian [1,2], Alexander Neugebauer [2], Lasse Scherffig [1] and Siegfried Wahl [2,3,*]

[1] Köln International School of Design, TH Köln, 50678 Köln, Germany; jkilian@kisd.de (J.K.); lscherff@kisd.de (L.S.)
[2] ZEISS Vision Science Laboratory, Eberhard-Karls-University Tübingen, 72076 Tübingen, Germany; a.neugebauer@uni-tuebingen.de
[3] Carl Zeiss Vision International GmbH, 73430 Aalen, Germany
* Correspondence: siegfried.wahl@uni-tuebingen.de; Tel.: +49-7071-29-84512

Abstract: This paper documents the design, implementation and evaluation of the Unfolding Space Glove—an open source sensory substitution device. It transmits the relative position and distance of nearby objects as vibratory stimuli to the back of the hand and thus enables blind people to haptically explore the depth of their surrounding space, assisting with navigation tasks such as object recognition and wayfinding. The prototype requires no external hardware, is highly portable, operates in all lighting conditions, and provides continuous and immediate feedback—all while being visually unobtrusive. Both blind (n = 8) and blindfolded sighted participants (n = 6) completed structured training and obstacle courses with both the prototype and a white long cane to allow performance comparisons to be drawn between them. The subjects quickly learned how to use the glove and successfully completed all of the trials, though still being slower with it than with the cane. Qualitative interviews revealed a high level of usability and user experience. Overall, the results indicate the general processability of spatial information through sensory substitution using haptic, vibrotactile interfaces. Further research would be required to evaluate the prototype's capabilities after extensive training and to derive a fully functional navigation aid from its features.

Keywords: tactile vision sensory substitution; blind; visually impaired; mobility; navigation; orientation; locomotion; open source; haptic; wearable

1. Introduction

1.1. Navigation Challenges for Blind People

Vision is the modality of human sensory perception with the highest information capacity [1,2]. Being born blind or losing one's sight later in life involves great challenges. The ability to cope with everyday life independently, to be mobile in unfamiliar places, to absorb information and, as a result, to participate equally in social, public and economic life can be severely hampered, ultimately affecting one's quality of life [3–6]. Society can, of course, address this issue on many levels, for example by ensuring accessibility of information or by designing public spaces to meet the specific needs of blind individuals or, more generally, visually impaired people (VIPs). In addition, technical aids, devices and apps are constantly being developed to assist VIPs with certain tasks [7,8]. These aids can essentially be divided into three aspects: obtaining information of the surroundings (What is written here? What is the object ahead like?), interfacing with machines and computers (input and output) and navigation (How do I get there? Can I walk this way?). While there are certainly large overlaps between these three aspects, this paper exclusively focuses on the third—navigation.

Following the definition of Montello and Sas [9], navigation (often used synonymously with mobility, orientation or wayfinding [10,11]) is the ability to perform "coordinated and goal-directed movement[s] through the environment". It can be divided into two subcomponents: *wayfinding* (long-range navigation to a target destination, spatial orientation, knowledge about surroundings and landmarks outside the immediate environment) and *locomotion* (short-range navigation, moving in the intended direction without colliding or getting stuck, obstacle detection and development of strategies to overcome them). This paper will concentrate on the latter and discuss how VIPs can be supported in this by technical aids—that is, without the help of human companions.

1.2. Ways to Assist Locomotion

One theoretical solution would be to rehabilitate (rudimentary) vision: since the 1990s, research has been conducted into surgical measures in both retinal implants stimulating the optic nerve and brain implants attached directly to the visual cortex. The quality of vision restored by this kind of surgery varies widely and, even in the best cases, represents only a fraction of the visual acuity of people with ordinary eyesight. Together with high costs, this leads to the fact that invasive measures of this kind are still far from widespread use and have so far only been tested on a small number of people. In the medium term, however, improving the quality of implants and simplifying surgical procedures could make the technology available to a wider public [12,13]. With regard to navigation, however, it is questionable whether the mere provision of visual information to the brain would adequately address the problem. Even if sighted individuals can rely on the visual modality for the most part, navigation and the acquisition of spatial knowledge required for it are by no means dedicated visual tasks. Blind people (but also sighted people to some extent) use multiple modalities and develop various strategies to master navigation and locomotion, which can be described as multimodal or even amodal task-specific functions. There is an ongoing discussion about how exactly VIPs—a very heterogeneous group with varying capacities for absorbing environmental information—obtain and cognitively process spatial knowledge [14,15].

Two well established and commercially available aids that do not attempt to assist locomotion through vision restoration alone are the white long cane (which provides basic spatial information about objects in close proximity) and the guide dog (which uses the dog's spatial knowledge to assist navigation).

Figures for prevalence of the *white cane* among VIPs vary greatly depending on the age of the study population, the severity of their visual impairment and other factors [16–19], but in one example (USA, 2008) it is as low as ~10% [16]. Even though the white cane, once mastered, has proven to be an extremely helpful (and probably the most popular) tool for VIPs, it comes with the drawback of having only a limited range (approx. 1 m radius) and not being able to recognise aerial obstacles (tree branches, protruding objects) [20]. Smart canes that give vibratory feedback, capable of recognising objects above waist height or further away, could offer a remedy, but have, to the best of our knowledge, not yet achieved widespread use. The reasons for this, as well as their advantages and disadvantages, have been discussed in various publications [20–24].

Guide dogs, on the other hand, are another promising option that can bring further advantages for blind people besides solving navigation tasks [25–28]. However, they are even less widespread (e.g., only 1% of the blind people in USA, 2008 [16] or 2.4% of "the registered blind" in the U.K. [26] owned one). The reasons for this and the drawbacks of the guide dog as a mobility aid are manifold and have been frequently discussed in the literature.

Even among the users of these two aids, 40% reported head injuries at least once a year (18% once a month) and as a result, 34% leave their usual routes only once or several times a month, while 6% never do [19].

With an estimated worldwide total number of 295 million people with moderate or severe visual impairment in 2020, of which 36 million are totally blind [29], it can be assumed that the locomotion navigation tasks addressed—among others—still represent a global problem for VIPs and pose an important field of research.

1.3. Sensory Substitution as an Approach

The concept of Sensory Substitution (SS) offers a promising approach to address this shortcoming. The basic assumption of SS is that the function of a missing or impaired sensory modality can be replaced by stimulating another sensory modality using the missing information. This only works because the brain is so plastic that it learns to associate the new stimuli with the missing modality, as long as they fundamentally share the same characteristics [30]. Surgical intervention would not be necessary because existing modalities or sensory organs can be used instead.

There is a great amount of scientific work on the topic of SS, specifically dealing with the substitution of visual information. The research field was established in the 1960s by a research group around Paul Bach-y-Rita; they developed multiple variations of a Sensory Substitution Device (SSD) stimulating the sense of touch in order to replace missing vision, commonly called Tactile-Vision Substitution Systems (TVSS) [31–33]. Many influential publications followed this example and developed SSDs for the tactile modality [34–37], one even suggesting using a glove as tactile interface [38]; others addressed further modalities, such as the auditory with so-called Auditory-Vision Substitution Systems (AVSS) [39–43]. While there are summaries of existing approaches [44,45], there are also many other, smaller publications on the topic in the literature—often only looking at a sub-area of SS.

It should be noted that there is further work on so-called Electronic Travel Aids (ETAs) [24,44,46]. SSDs differ from them in the sense that they pass largely unprocessed information from the environment on to the substituting sensory modality and leave the interpretation to the user, while ETAs provide pre-interpreted abstract information (e.g., when to turn right or where a certain object is located).

Additionally, there is also research on SS-like assistive devices outside the scientific context [47–50].

1.4. Brain Plasticity and Sensory Substitution

The theoretical basis of SS is summarised under the term of brain plasticity. Although the focus of this paper is not on the neurophysiological discussion of this term, a brief digression is nevertheless helpful in order to understand some of the design decisions made in this project.

In general, *brain plasticity* describes the "adaptive capacities of the central nervous system" and "its ability to modify its own structural organization and functioning" [30]. While neuroscience has long assumed a fixed assignment of certain sensory and motor functions to specific areas of the brain, we today know that the brain is capable of reorganising itself, e.g., after brain damage [51] and, moreover, is capable of learning new sensory stimuli not only in early development but throughout life [52]. For sensory substitution to work and for the new neural correlate to be learned, a number of conditions are nevertheless necessary; Bach-Y-Rita et al. [34] point out that there has to be

> "(a) functional demand, (b) the sensor technology to fill that demand, and (c) the training and psychosocial factors that support the functional demand".

An SSD only needs a sensor that picks up the information, an interface that transmits it to human receptors, and finally and very importantly, the possibility for the user to modify the sensory stimulus by motor action in order to determine the initial origin of the information [34]. The importance of the latter close dependence between motor action and sensory perception has been emphasised in many publications [33,34] and is assumed to be the basis of *any* sensory experience [53].

There still is a vital discussion across disciplines about how cognitive processing of sensory stimuli is carried out by the brain. Worth mentioning here is the Theory of Sensorimotor Contingencies (SMCs) dismissing longstanding representational models and describing the supposedly passive perception of environmental cues as an active process that relies on regularities between action and reception that have to be learned [54]. The literature on SSDs and the SMC theory mutually refer to each other [54,55].

1.5. Pitfalls of Existing Substitution Systems

Despite the long tradition of research on the topic of SS and numerous publications with promising results, the concept has not yet achieved a real breakthrough. The exact reasons for the low number of available devices and users have often been discussed and made the subject of proposals for improvement [30,40,44,55,56].

Certain prerequisites that an SSD must meet in order to be used by a target group can be gathered from both existing literature and methods of interaction design. These aspects are of a very abstract nature and their implementation in practice is challenging and often only partially achievable. The following 14 aspects or prerequisites—the first ten of which were originally formulated as *problems* of existing SSDs by Chebat et al. [55]—were taken into account in the design and evaluation of the proposed SSD: learning, training, latency, dissemination, cognitive load, orientation of the sensor, spatial depth, contrast, resolution, costs, motor potential, preservation of sensory and motor habits, user experience and joy of use, and aesthetic appearance. See Appendix A for a discussion.

The motivation to deal with this field arose from the technological progress since Bach-y-Rita's early pioneering work and his analogue experimental set-ups; improvements in price, portability and processing power of modern digital computing devices, sensors and other components necessary for the development of an SSD have opened up many new possibilities in the field and facilitate the implementation of these 14 aspects, now more than ever. Recent literature on assistive technology for VIPs also suggests that the field is "gaining increasing prominence owing to an explosion of new interest in it from disparate disciplines" having a "very relevant social impact" [57].

2. The Unfolding Space Glove

2.1. Previous Work

The Unfolding Space Glove is an SSD; it translates spatio-visual depth images into information that can be sensed haptically. In a broader sense, it is a TVSS, with the original term *tactile* perception deliberately being changed to *haptic* because of the high level of integration of the motor system, and the term *visual* being changed to *spatio-visual* to describe the input more accurately.

It was first drafted in previous work by Kilian in 2018 [58,59] with a focus on Interaction Design (only available in German). However, the first prototypes of the glove were still a bit cumbersome, heavy, had higher latencies and were prone to errors. Nevertheless, they were able to prove the functional principle and demonstrate that more research on this evice is worthwhile. Through the course of this project, the device was refined and the ototype tested in this study was ultimately able to deliver promising results in the em- cal tests (see results and discussion section) and meets many of the previously defined equisites (see discussion section). For more details and background information on the ct please also see the project website https://www.unfoldingspace.org (accessed on 26 y 2022). Code, hardware, documentation and building instructions are open source ailable in the public repository https://github.com/jakobkilian/unfolding-space d on 26 January 2022). Consider Release v0.2.1 for the stable version used in this d consider more recent commits in which the content has been revised for better ty.

e and Technical Details

olding Space Glove (Figure 1) essentially consists of two parts: a USB power orn on the upper arm (or elsewhere) and the glove itself, holding the camera, puting unit and associated technology. The only connection between them pply USB cable. A list of the required components and photographic f the assembly process is attached in the Supplementary Materials S1 and aforementioned Github repository. The mere material costs of the entire 00, of which about two thirds go to the camera alone (see Appendix B).

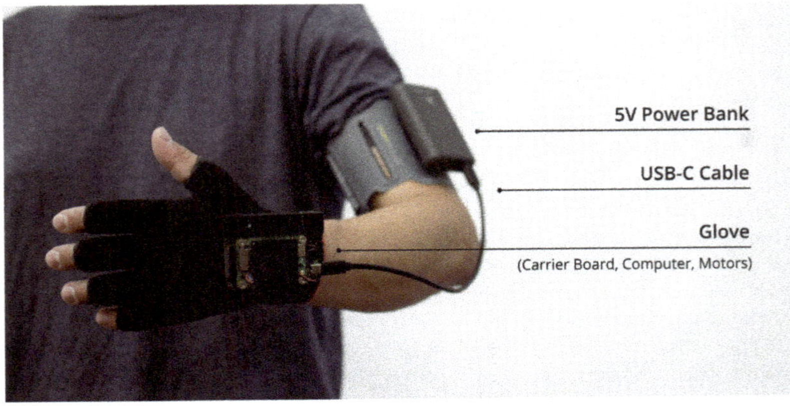

Figure 1. The final prototype with labeled components.

Structurally, a TVSS typically consists of three components: the input (a camera that captures images of the environment), the computing unit (translating these images into patterns suitable for tactile perception), and finally the output (a tactile interface that provides this information to the user).

The selected input system gathering 3D information on the environment is the "Pico Flexx" Time of Flight (ToF) camera by *pmdtechnologies* [60]. ToF refers to a technique that determines the distance to objects by measuring the time that actively emitted light signals take to reach them (and bounce back to the sensor). For quite a while, SS research focused on conventional two-dimensional greyscale images, and it is only in the last few years that the use of 3D images or data, especially from ToF cameras, has been investigated. Due to the advantages of today's ToF technology compared to other 3D imaging methods (such as structured light or stereo vision) [61], it seemed reasonable to make efforts to explore exactly this combination. See Appendix B for a more detailed discussion of existing 3D SSDs and the choice of the ToF camera technology.

The computing unit, a Raspberry Pi Compute Module 4, is attached to the glove as part of the *Unfolding Space Carrier Board* which is described more closely in Appendix C.

The output is a 3 × 3 matrix of (vibrating) linear resonant actuators (LRAs) placed on the back of the hand in the glove. The choice of actuators and the reasons for positioning them on the back of the hand are described in Appendix D.

2.3. Algorithm

The SSD's translation algorithm takes the total of 224 × 171 (depth) pixels from the ToF camera and calculates the final values of the glove's 3 × 3 vibration pattern. Each motor represents the object that is closest to the glove within the corresponding part of the field of view of the camera. A detailed explanation of the algorithm can be found in Appendix E. The code files and their documentation are attached as zip files in the Supplementary Materials S2 but can also be accessed in the aforementioned Github repository.

2.4. Summary

The resulting system achieves a frame rate of 25 fps and a measured latency of about 50 ms. About 10 ms of this is due to the rise time of the LRAs, 3 ms to the image-processing on the Raspberry Pi and an unknown part to the operations of the Pico Flexx specific library. The system has a horizontal field of view of 62° and a vertical of 45° with a detection range from 0.1 m to 2 m [60]. The beginning of the detection range is determined by the limitations of the ToF method, while the 2 m maximum distance was just fixed for this study and could be adjusted in the future (maximum of the camera is 4 m [60] with decreasing quality in the far range). The glove by itself weighs 120 g with all its components, the power bank and bracelet weigh 275 g together, giving a total system weight of 395 g. The glove was produced in two sizes, each with a spare unit, in order to guarantee a good fit for all subjects and a smooth conduct of the study.

Now that the physical system has been described in detail, the next section will explain the study design.

3. Experimental Methods

3.1. Ethics

In order to evaluate the SSD described, a quasi-experiment was proposed and approved by the ethics committee of the faculty of medicine at the university hospital of the Eberhard-Karls-University Tübingen in accordance with the 2013 Helsinki Declaration. All participants were informed about the study objectives, design and associated risks and signed an informed consent form to publish pseudonymous case details. The individuals shown in photographs in this paper have explicitly consented to the publication of those by signing an additional informed consent form.

3.2. Hypotheses

The study included training and testing of the white long cane. This was not done with the intention of pitting the two aids against each other or eventually replacing the cane with the SSD. Rather, the aim was to be able to discuss the SSD comparatively with respect to a controlled set of navigation tasks. In fact, the glove is designed in a way that it could be worn and used in combination with the cane in the other hand. Testing not only both aids but also the combination of both would, however, introduce new unknown interactions and confounding factors. The main objective of this study thus reads: "The impact of the studied SSD on the performance of the population in both navigation and obstacle detection is comparable to that of the white long cane." The hypothesis derived from this is complex due to one problem: blind subjects usually have at least basic experience with the white cane or have been using it on a daily basis for decades. A newly learned tool such as the SSD can therefore hardly be experimentally compared with an internalised device like the cane. A second group of naive sighted subjects was therefore included to test two separate sub-hypotheses:

Hypothesis 1. ***Performance.***

Sub-Hypothesis H1a. ***Non-Inferiority of SSD Performance:*** *after equivalent structured training with both aids, sighted subjects (no visual impairment but blindfolded, no experience with the cane) achieve a non-inferior performance in task completion time (25 percentage points margin) with the SSD compared to the cane in navigating an obstacle course.*

Sub-Hypothesis H1b. ***Equivalency of Learning Progress across Groups:*** *at the same time, blind subjects who have received identical training (here only with the SSD) show equivalent learning progress with the SSD (25 percentage points margin) as the sighted group.*

With both sub-hypotheses confirmed, one can therefore, in simple terms, make assumptions about the effect of the SSD on the navigation of blind people compared to the cane, if both had been learned similarly. In addition, two secondary aspects should be investigated:

Hypothesis 2. ***Usability & Acceptance.*** *The device is easy to learn, simple to use, achieves a high level of user enjoyment and satisfaction and thus strong acceptance rates.*

Hypothesis 3. ***Distal Attribution.*** *Users report unconscious processing of stimuli and describe the origin of these haptic stimuli distally in space at the actual location of the observed object.*

3.3. Study Population

A total of 14 participants were recruited mainly through calls at the university and through local associations for blind and visually impaired people in Cologne, Germany. Appendix F contains a summary table with the subject data presented below. The complete data set is also available in the study data in Supplementary Materials S3.

Six of the subjects were normally sighted and had a visual acuity of 0.3 or higher; eight were blind (congenitally and late blind), thus had a visual acuity of less than 0.05 and/or a visual field of less than 10° (category 3–5 according to ICD-10 H54.9 definition) on the better eye. Participants' self-reports about their visual acuity were confirmed with a finger counting test (1 m distance) and, if passed, with the screen based Landolt visual acuity test "FrACT" [62] (3 m distance) using a tactile input device (Figure 2A).

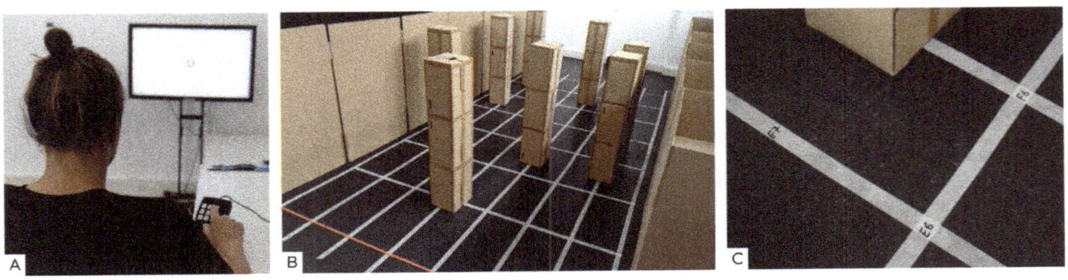

Figure 2. (**A**) Carrying out the "FrACT" Landolt visual acuity test with screen and tactile input device (lower right corner). (**B**) The obstacle course in the study room. (**C**) Close up of the grid system used for quick rebuilding of the layouts.

Two subjects were excluded from the evaluation despite having completed the study: on average, subject f (cane = 0.75, SSD = 2.393) and subject z (cane = 1.214, SSD = 1.500) caused a remarkably higher number of contacts (two to three-fold) with *both* aids than the average of the remaining blind subjects (cane = 0.375, SSD = 0.643). For the former this can be explained by a consistently high level of nervousness when walking through the course. With both aids, the subject changed their course very erratically in the event of a contact, causing further contacts or even collisions right away. The performance of the latter worsened considerably towards the end of the study, again with both aids, so much so that the subject was no longer able to fulfil the task of avoiding obstacles at all, citing "bad form on the day" and fatigue as reasons. In order to not influence the available data by this apparent deviation, these two subjects were excluded from all further analysis.

The age of the remaining participants (six female, five male, one not specified) averaged 45 ± 16.65 years and ranged from 25 to 72 years. All were healthy—apart from visual impairments—and stated that they were able to assess and perform the physical effort of the task; none had prior experience with the Unfolding Space Glove or other visual SSDs.

All participants in the blind group have been using the white cane on a daily basis and for at least five years and/or did an Orientation and Mobility (O&M) training. Some occasionally use technical aids like GPS-based navigation and one even had prior experience using the *feelspace* belt (for navigation reasons only, not for augmentation of the Earth's magnetic field). Two reported to use *Blind Square* from time to time and one used a monocular.

None of the sighted group had prior experience with the white cane.

3.4. Experimental Setup

The total duration of the study per subject differs between the blind and the sighted group, as the sighted have to do the training with both aids and the blind with the SSD only (since one inclusion criterion was experience in using the cane). The total length thus was about 4.5 h in the blind group and 5.5 h in the sighted group.

In addition to paper work, introduction and breaks, participants of the sighted group received 10 min of an introductory tutorial on both aids, had 45 min of training with them, spent 60 min using them during the trials (varied slightly due to the time required for completion) and thus reached a total wearing time of about 2 h with each aid. In the blind group, the wearing time of the SSD was identical, while the wearing time of the cane is lower due to the absence of tutorial and training sessions with it.

The study was divided into three study sessions, which took place at the Köln International School of Design (TH Köln, Cologne, Germany) over the span of six weeks. In the middle of a 130 square meter room, a 4 m wide and 7 m long obstacle course was built (Figure 2B), bordered by 1.80 m high cardboard side walls and equipped with eight cardboard obstacles (35 × 31 × 171 cm) distributed on a 50 cm grid (Figure 2C) according to the predefined course layouts.

3.5. Procedure

Before the first test run (baseline), the participants received a 10-min *Tutorial* in which they were introduced to the handling of the devices. Directly afterwards, they had to complete the first *Trial Session (TS)*. This was followed by total of three *Practices Sessions (PS)*, each of them being followed by another TS—making the Tutorial, four TS and three PS in total. The study concluded with a questionnaire at the end of the third study session after completion of the fourth and very last TS. An exemplary timetable of the study procedure can be found in the Supplementary Materials S4.

3.5.1. Tutorial

In the 10-min Tutorial, participants were introduced to the functional design of the device, its components and its basic usage such as body posture and movements while interacting with the device. At the end, the participants had the opportunity to experience one of the obstacles with the aid and to walk through a gap between two of these obstacles.

3.5.2. Trial Sessions

Each TS consisted of seven consecutive runs in the aid condition cane and seven runs in the condition SSD, with a flip of a coin in each TS deciding which condition to start with. The task given verbally after the description of the obstacle course read:

> "You are one meter from the start line. You are not centered, but start from an unknown position. Your task is to cross the finish line seven meters behind the start line by using the aid. There are eight obstacles on the way which you should not touch. The time required for the run is measured and your contacts with the objects are counted. Contacts caused by the hand controlling the aid are not counted. Time and contacts are equally weighted—do not solely focus on one. You are welcome to think out loud and comment on your decisions, but you won't get assistance with finishing the task."

Contacts with the cane were not included in the statistics, as an essential aspect of its operation is the deliberate induction of contact with obstacles. In addition, for both aids, contacts caused by the hand guiding it were not included in the statistics as well in order to motivate the subjects to freely interact with the aids. There was a clicking sound positioned at the end of the course (centred and 2 m behind the finish line) to roughly guide the direction. There was no help or other type of interference while participants were performing the courses. Only when they accidentally turned more than 90 degrees away from the finish line were they reminded to pay attention to the origin of the clicking sound. Both task completion time and obstacle contacts (including a rating in mild/severe contacts) were entered into a macro-assisted Excel spreadsheet on a hand-held tablet by the experimenter, who was following the subjects at a non-distracting distance. The data of all runs can be found in the study data in Supplementary Materials S3.

A total of 14 different course layouts were used (Figure 3), seven of which were longitudinal axis mirror images of the other seven. The layout order within one aid condition (SSD/cane) over all TS was the same for all participants and predetermined by drawing all 14 possible variations for each TS without laying back. This means that all participants went through the 14 layouts four times each, but in a different order for each TS and with varying aids, so that a memory effect can be excluded.

The layouts were created in advance using an algorithm that distributed the obstacles over the 50 cm grid. A sequence of 20 of these layouts was then evaluated in self-tests and with pre-subjects, leaving the final seven equally difficult layouts (Figure 3).

The study design and the experimental setup were inspired by a proposal of a standardised obstacle course for assessment of "visual function in ultra low vision and artificial vision" [63] but has been adapted due to spatial constraints and selected study objectives (e.g., testing with two groups and limited task scope only). There are two further studies suggesting a very similar setup for testing sensory substitution devices [64,65] that were not considered for the choice of this study design.

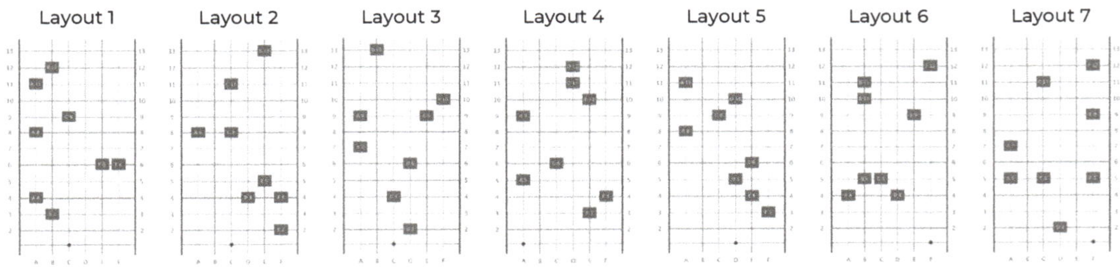

Figure 3. The seven obstacle course layouts used in the study (in non-mirrored variant).

3.5.3. Practice Session

The practice sessions were limited to 15 min and followed a fixed sequence of topics and interaction patterns to be learned with the two aids (Figure 4A–C). In the training sessions obstacles were arranged in varying patterns by the experimenter. Subjects received support as needed from the experimenter and were not only allowed to touch objects in their surroundings, but were even encouraged to do so in order to compare the stimuli perceived by the aid with reality.

In the case of the SSD training, after initially learning the body posture and movement, the main objective was to understand exactly this relationship between stimuli and real object. For this purpose, the subjects went through, for example, increasingly narrow passages with the aim of maintaining a safe distance to the obstacles on the left and right. Later, the tasks increasingly focused on finding strategies to find ways through course layouts similar to the training layouts.

Figure 4. (**A**) A subject practising with the SSD in the tutorial at the beginning of the study (**B**) A typical task in the practice sessions: passing a gap between two obstacles (**C**) A sighted subject training with the cane during a practice session.

While the training with the white cane (in the sighted group) took place in comparable spatial settings, here the subjects learned exercises from the cane programme within an O&M training (posture, swing of the cane, gait, etc.). The experimenter himself received a basic cane training from an O&M trainer in order to be able to carry it out in the study. The sighted subjects were therefore not trained by an experienced trainer, but all by the same person. At the same time, the SSD was not trained "professionally" either, as there are no standardised training methods specifically for the device yet.

3.5.4. Qualitative Approaches

In addition to the quantitative measurements, the subjects were asked to think aloud, to comment on their actions and to describe why they made certain decisions, both during training and during breaks between trials. These statements were written down by hand by the experimenter.

After completion of the last trial, the subjects were asked to fill out the final questionnaire. It consisted of three parts: firstly, the 10 statements of the System Usability Score (SUS) Test [66] on a 0–4 Likert agreement scale; Secondly, 10 further custom statements on handling and usability on the same 0–4 Likert scale; And finally seven questions on different scales and in free text about perception, suggestions for improvement and the possibility to leave a comment on the study. The questions of part one and two were always asked twice: once for the SSD and once for the cane. The subjects could complete this part of the questionnaire either by handwriting or with the help of an audio survey with haptic keys on a computer. This allowed both sighted and blind subjects to answer the questions without being influenced by the presence of the investigator. The third part, on the other hand, was read out to the blind subjects by the investigator, who noted down the answers.

Due to the small number of participants, the results of the questionnaire are not suitable for drawing statistically significant conclusions, but should rather serve the qualitative comparison of the SSD with the cane and support further developments on this or similar SSDs. In the study data in Supplementary Materials S3 there is a list with all questionnaire items and the Likert scale answers of items 1–20. In the Results section and in Appendix H relevant statements made in the free text questions are included.

3.6. Analysis and Statistical Methods

A total of 784 trials in 14 different obstacle course layouts were performed by every subject over all sessions. The dependent variables were *task completion time* (in short *time*) and *number of contacts* (in short *contacts*).

Fixed effects were:

- *group:* between-subject, binary (blind/sighted)
- *aid:* within-subject, binary (SSD/cane)
- *TS:* within-subject, numerical and discrete (the four levels of training)

Variables with random effects were: *layout* as well as the *subjects* themselves, nested within their corresponding group.

The quantitative data of the dependent variable task completion time was analysed by means of parametric statistics using a linear mixed model (LMM). In order to check whether the chosen model corresponds to the established assumptions for parametric tests, the data were analysed according to the recommendation of Zuur et al. [67]. The time variable itself has been normalised in advance using a logarithmic function to meet those assumptions (referred to in the following as *log time*). With the assumptions met, all variables were then tested for their significance to the model and their interactions with each other. See Appendix G for details on the model, its fitting procedure, the assumption and interactions tests and corresponding plots.

Most statistical methods only test for the presence of differences between two treatments and not for their degree of similarity. To test the sub-hypotheses of H1, a non-inferiority test (H1a) and an equivalence test (H1b) were thus carried out. These check whether the least squares (LS) means and corresponding confidence intervals (CI) of a selected contrast exceed a given range (here 25 percentage points in both sub-hypotheses) either in the lower or in the upper direction. In order to confirm the latter, equivalence, both directions must be significant; For non-inferiority only the "worse" (in this case the slower side) has to be significant (since it would not falsify the sub-hypothesis if the SSD were unequally faster) [68].

No statistical tests were performed on the contacts data. Since the data structure is zero-inflated and poisson distributed, non-parametric tests such as a generalised linear mixed model would be required, resulting in low statistical power given the sample size. Nevertheless, descriptive statistical plots of these data alongside the analysis of the log time statistics are to be included in the next section.

All analyses have been executed using the statistical computing environment *R* and the graphical user interface *RStudio*. The *lme4* package was used to run LMMs. To calculate Least LS means, their CI and the non-inferiority/equivalency tests, the *emmeans* and the *emtrends* package was used. In this paper averages are shown as arithmetic mean with the corresponding standard deviation. For all statistical tests an alpha level of 0.05 was chosen.

4. Results

4.1. Overview

To give an impression of the study procedure, a series of videos was made available in high resolution at https://vimeo.com/channels/unfoldingspace (accessed on 26 January 2022). A selection of lower resolution clips is also attached to Supplementary Materials S5. The corresponding subject identifier and the trial number can be found in the opening credits and the descriptive text. All test persons shown here have explicitly agreed to the publication of these recordings.

To get an overview of the gathered data, Figure 5 shows log time (the normalised time) for both aid conditions over the four TS horizontally split by group. The plot contains all individual samples as dots in the background, summarised in boxplots including the mean (marked with the "×" sign) in addition to the common median line. The solid line represents the linear regression of the fitted LMM. In this and in following plots the variables aid and group are shown as combination with the levels sighted & SSD (S&S), sighted & cane (S&C), blind & SSD (B&S) and blind & cane (B&C).

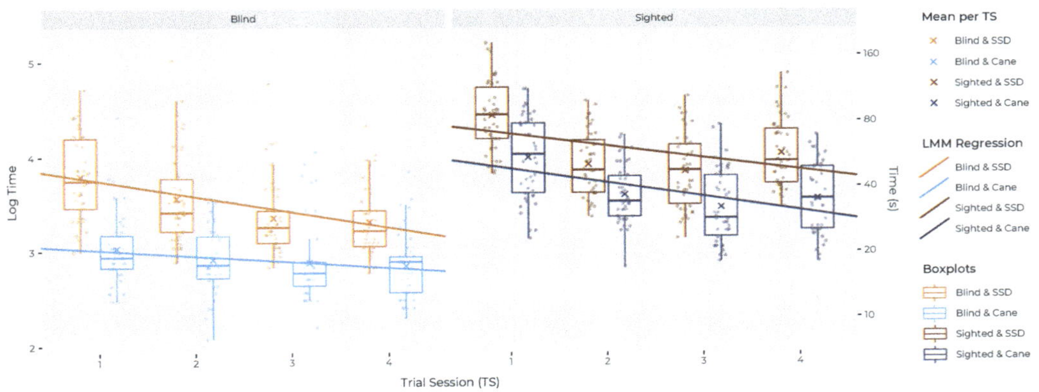

Figure 5. Trial session (TS) vs. normalised task completion time (log time) of all samples as boxplots, means and regression lines from the linear mixed model. Split horizontally by group (blind/sighted) and subdivided by aid in colour (SSD/cane).

Contacts show a similar picture (Figure 6): in the last TS, sighted subjects touched an average of 0.38 ± 0.73 objects per run with the SSD and only 0.12 ± 0.33 with the cane. Blind subjects also showed a comparable response in the last TS, touching an average of

0.45 ± 0.89 objects per run with the SSD, while touching only an average of 0.4 ± 0.63 objects with the cane. As mentioned, these differences cannot be reasonably tested.

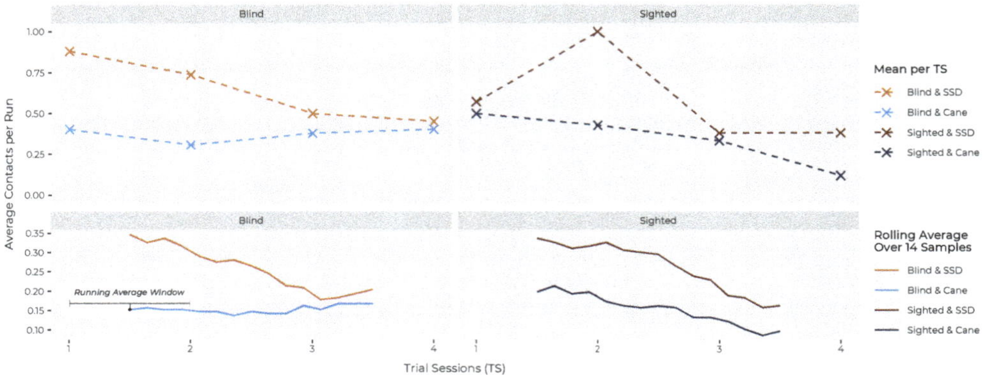

Figure 6. Trial sessions (TS) vs. average number of contacts with obstacles per run. The upper half shows the means per TS; The lower one shows a running average (running over all samples) calculated to better visualise trends. Note the wide window of 14 samples per data point. The plot is split horizontally by group (blind/sighted) and subdivided by aid in colour (SSD/cane).

4.2. Learning Effect

As a basic assumption for the subsequent hypothesis tests, the learning effect on performance has to be investigated. A significant effect of TS on log time was expected in all combinations except B&C, in which the subjects were already familiar with the aid. Still, with habituation to the task and familiarity with the conditions of the course (size, type of obstacles, etc.), a negligible effect could be expected in all four conditions, i.e., also in the case of B&C. The test was carried out by adjusting the base level of the LMM to the four different combinations. TS shows a statistically significant effect on log time in Condition S&S (intercept = 4.37, slope = −0.13, SE = 0.04 , $p = 0.007$), S&C (intercept = 4.03, slope = −0.15, SE = 0.04 , $p = 0.002$) and in B&S (intercept = 3.91, slope = −0.16, SE = 0.04 , $p = 0.001$) but *not* in B&C (intercept = 3.08, slope = −0.06, SE = 0.04 , $p = 0.137$). The expected learning progress was thus confirmed by the tests; the general habituation slope over the course of the study for all subjects and aids was around −0.06 s on log scale.

4.3. Hypothesis H1 | Performance

Given the knowledge of the significant effects of group, aid and TS and their interactions, the two sub-hypothesis H1a and H1b could be tested.

4.3.1. H1a | Non-Inferiority of SSD Performance

In order to accept H1a, two separate tests were carried out: *firstly*, a pairwise comparison using Tukey's HSD test between conditions S&S and S&C—both under the condition of TS being 4 (after last training): using the Kenward–Roger approximation, a significant difference ($p < 0.001$) was found, with the log time LS means predicted to be 54.43% slower with the SSD (3.87 s, SE = 0.13) than with the cane (3.26 s, SE = 0.13). Back transformed to the response scale, this gives a predicted time of 47.9 s (95% CI [64.2 s, 35.8 s]) for the SSD and 31.0 s (95% CI [41.6 s, 23.2 s]) for the cane to complete one obstacle course run. *Secondly*, the test for non-inferiority between these two conditions (using the Kenward–Roger approximation and the Šidák correction) was performed and found to be non-significant (p value 1). This means that the SSD is significantly different from the cane and could be considered inferior within the predefined tolerance range of 25%. H0 of H1a thus *could not be rejected*. The difference between SSD and cane under the condition investigated can also be observed in Figure 7A.

For contacts, as mentioned, a statistical analysis is not feasible. Still, the results can be compared descriptively in previous plot Figure 6. As already mentioned, the difference in measured mean contacts per run differed from 0.38 ± 0.73 objects per run with the SSD and only 0.12 ± 0.33 with the cane.

4.3.2. H1b | Equivalence of Learning Progress

To accept H1b, the learning progress of the SSD had to be compared between the two groups, again by using two tests: *firstly*, the estimated effect of TS on log time differed by only 0.04 s (SE = 0.06) between S&S (−0.12 s, SE = 0.04) and B&S (−0.16 s, SE = 0.04) condition, while not being significant (p value = 0.89). This means that there is no proof at this point that the learning progress between the groups is different. *Secondly*, to examine the degree of similarity of the given contrast an equivalence test was carried out (again using Kenward–Roger approximation and Šidák correction): a significant p value of 0.016 indicated the presence of equivalence of learning progress in both groups with the SSD (within the predefined tolerance range of 25%). H0 of H1b thus *could be rejected*. The learning progress of the SSD across the groups can also be observed in Figure 7B.

Again for contacts, a statistical analysis is not feasible. Nevertheless, it appears useful to compare the progress of the sighted and blind curve with the SSD in previous plot Figure 6. In particular, the running average described in this figure suggested a quite similar progress between those two.

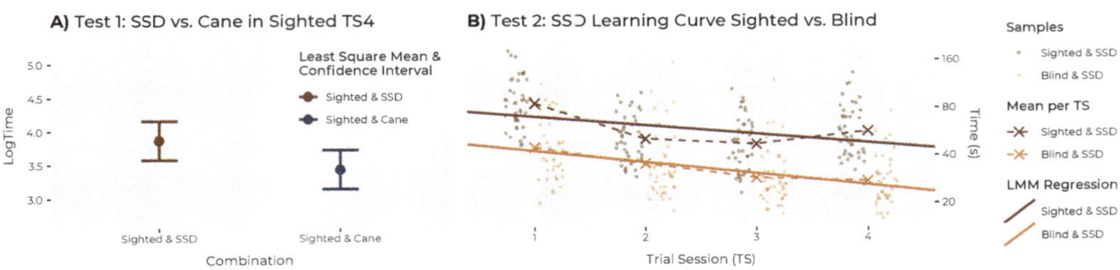

Figure 7. Graphical representations of the two main hypothesis tests. (**A**) Sub-hypothesis H1a: performance with the two devices (SSD/cane) after the last training (only Trial Session 4). Least square means and the corresponding confidence interval extracted from the linear mixed model are shown to demonstrate the test. (**B**) Sub-hypothesis H1b: the learning curves with the SSD in both groups (sighted/blind) over the course of the training (all Trial Sessions) are shown. The test only compares the linear regression slopes, but means and the underlying data points are plotted here as well.

4.4. *Hypothesis H2 | Usability & Acceptance*

In Figure 8, one can find a tabular evaluation and a graphical representation of all Likert scale questions of the first and second part of the questionnaire (including the SUS). In general, one can see that the degree of coverage between SSD and cane was comparatively high. The discussion section therefore looks at the questions that show the greatest average deviations and discusses them in a classifying manner. There is no graphical representation of questions 21–27 as they were in free text or on other scales.

4.4.1. System Usability Score

The System Usability Score, which was queried in the first 10 questionnaire items, results from the addition of all scores multiplied by 2.5 and thus ranges from 0 to a maximum of 100 possible points. The SSD achieved an average SUS of 50 in this study, while the cane scored quite similarly at 53. As expected, the cane performed slightly better in the blind group (54) than in the sighted (52), while the SSD performed better in the sighted (51) than in the blind (49). The differences are rather negligible due to the sample size but can be seen as an indicator of a quite comparable assessment of both systems.

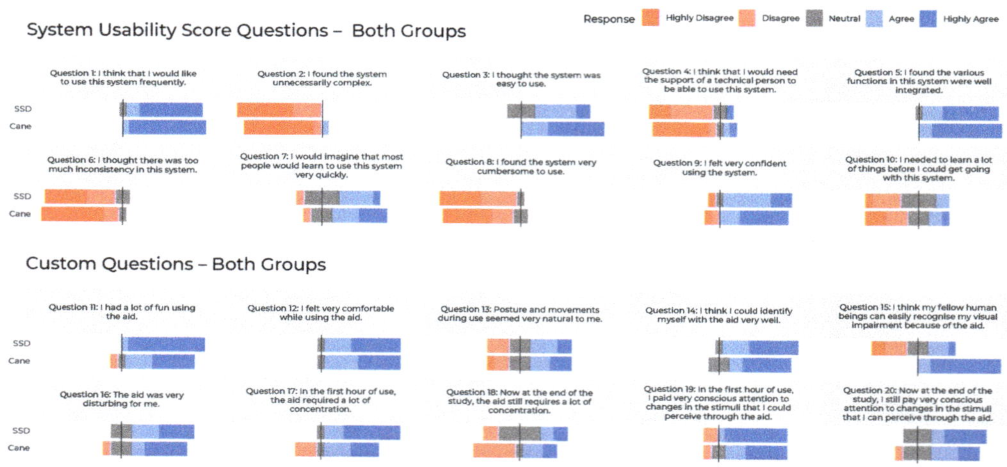

Figure 8. A summary of all Likert scale questionnaire items. The results are summarised for both groups (sighted/blind) and split per question according to SSD and cane.

4.4.2. Clustered Topics in Subjects' Statements

Statements expressed in the free interviews after each session, during the aloud reflection in the training sessions and in the free text questions of the questionnaires were grouped into five main topics (with the number of subjects mentioning them in brackets):

- **Cognitive Processing of the Stimuli (9).** From the statements it is quite clear that the use of the SSD at the beginning of the study required a considerably higher cognitive effort than the use of the cane (Appendix H, Table A3, ID 1–4). Towards the end of the study, the subjects still reported a noticeable cognitive effort. However, they often also noted that the experience felt different than it did at the beginning of the training and that they could imagine that further training would reduce the effort even more (Appendix H, Table A3, ID 5 & 6). The subjects' reports towards the end of the study also suggested that deeper and more far-reaching experiences might be possible with the SSD than with the cane (Appendix H, Table A3, ID 7–9).
- **Perception of Space and Materiality (6).** The topic of how subjects perceived space and its materiality is undoubtedly related to the previously cited statements about the processing of stimuli: it is noteworthy how often spatial or sometimes visual accounts were assigned to the experiences with the SSD, while the cane was rather described as a tool for warning of objects ahead (Appendix H, Table A4).
- **Wayfinding Processes (5).** It was mentioned as an advantage of the glove that, in contrast to the cane, an obstacle can already be detected before contact is made—i.e., earlier and from a greater distance; A different path can then be taken in advance in order to avoid a collision with this object. In addition, some described an unpleasant feeling of actively bumping into obstacles with the cane just to get information. However, these were mainly sighted people who were not yet used to handling the cane (Appendix H, Table A5).
- **Enjoyment of Use (3).** The cane is described by some as easy to learn but therefore less challenging and less fun (Appendix H, Table A6).
- **Feeling Safe and Comfortable (3).** On the other hand, subjects also report that they feel safer and more comfortable with the cane (Appendix H, Table A7).

4.4.3. Advantages of the SSD

In question 25 (Q25) the subjects could name pros and cons of both devices. These were summarised to topics, being described from the perspective of the advantages of the Unfolding Space Glove. An advantage of the cane, for example, was thus evaluated as a disadvantage of the SSD. The most frequently mentioned (by three or four subjects) advantages of the SSD were the following: more spatial awareness is possible; one can survey a wider distance; the handling is more subtle and quiet. Frequent disadvantages were: the higher learning effort for the SSD and the fact that one can obtain less information about the type of objects due to missing acoustic feedback.

4.4.4. Suggestions for Improvement from Subjects

In Q26, the subjects were encouraged to list their suggestions for improvement for a future version of the same SSD—even if these may not be technically feasible. The two biggest wishes addressed two well-known problems of the prototype: detection of objects close to the ground (e.g., steps, thresholds, unevenness, …) was requested by five subjects and the detection of objects closer than 10 cm (where the prototype currently cannot measure and display anything) was requested by four of them. Both would probably have been mentioned even more frequently if they had not already been pointed out as well-known problems at the beginning of the study. Additionally, subjects (number in brackets) wished that they did not have to wear a battery on their arm (3) and wished that the device was generally more comfortable (2). Some individuals mentioned that they would like to customise the configuration (e.g., adjust the range). Some wished for the detection of certain objects (e.g., stairs) or characteristics of the room (brightness/darkness) to be communicated to them via vibration patterns or voice output.

4.5. Hypothesis H3 | Distal Attribution

H3 could be rejected, as there were no specific indications of distal attribution of perceptions in the subjects' statements. However, some of the statements strongly suggest that such patterns were already developing in some subjects (Appendix H, Table A8), which is why this topic will be addressed in the discussion.

5. Discussion

The results presented above demonstrate not only the perceptibility and processibility of 3D images by means of vibrotactile interfaces for the purpose of navigation, but also the feasibility, learnability and usefulness of the novel Unfolding Space Glove—a haptic spatio-visual sensory substitution system.

Before discussing the results, it has to be made explicitly clear that the study design and the experimental set-up do not yet allow generalisations to be made about real-life navigational tasks for blind people. In order to be able to define the objective of the study precisely, many typical, everyday hazards and problems were deliberately excluded. These include objects close to the ground (thresholds, tripping hazards and steps) or the recognition of approaching staircases. Furthermore, auditory feedback from the cane, which allows conclusions to be drawn about the material and condition of the objects in question, were omitted. In addition, there is the risk of a technical failure or error, the limit of a single battery charge and other smaller everyday drawbacks (waterproofness, robustness, etc.) that the prototype currently still suffers from. Of course, many of the points listed here could be solved technically and could be integrated into the SSD at a later stage. However, they would require development time, would have to be evaluated separately and can therefore not simply be taken for granted in the present state.

With that being said, it is possible to draw a number of conclusions from the data presented. First of all, some technical aspects: the prototype withstood the entire course of the study with no technical problems and was able to meet the requirements placed on it that allow the sensory experience itself to be assessed as independently of the device as possible. These include, for example, intuitive operation of the available functions, sufficient wearing comfort, easy and quick donning and doffing, sufficient battery life and good heat management.

The experimental design can also be pointed out: components such as data collection via tablet, the labelled grid system for placing the obstacles plus corresponding set-up index cards and the interface for real-time monitoring of the prototype enabled the sessions to be carried out smoothly with only one experimenter. An assistant helped to set up and dismantle the room, provided additional support (e.g., by reconfiguring the courses and documented the study in photos and videos), but neither had to be, nor was present at every session. Observations, ratings and participant communication were carried out exclusively by the experimenter.

Turning now to the sensory experience under study, it can be deduced that 3D information of the environment is very direct and easy to learn. Not only were the subjects able to successfully complete all 392 trials with the SSD, but they also showed good results as soon as the first session and thus after only a few minutes of wearing the device. This is, to the best of our knowledge, in contrast to many other SSDs in the literature, which require several hours of training before the new sensory information can be used meaningfully (in return, usually offering higher information density).

Nevertheless, the cane outperforms the SSD in time and contacts, in both groups in the first TS and at every other stage of the study (also see Figures 5 and 6). Apart from the measurements, the fact that the cane seems to be even easier to access than the glove is also shown in the results of the questionnaire among the sighted subjects, for whom both aids were new: while many answers between the SSD and the cane only differed slightly on average ($\Delta \leq 0.5$), the deviations are greatest in questions about the learning progress. Sighted subjects thought the cane was easier to use (Q3, $\Delta = 1.0$), could imagine that "most people would learn to use this system" more quickly (Q7, $\Delta = 0.8$) and stated that they had to learn less things before they could use the system (Q10, $\Delta = 0.7$) compared to the SSD.

5.1. Hypothesis 1 and Further Learning Progress

At the end of the study and after about 2 h of wearing time, H1a states that the SSD is still about 54% slower than the cane. Even though this difference in walking speed could be acceptable if (and only if) other factors gave the SSD an advantage, H1 had to be rejected: the deviation exceeds the predefined 25% tolerance range and thus can no longer be understood as an "non-inferior performance". H1b, however, can be accepted, indicating that due to a "equivalent learning progress" between the groups these results would also apply to blind people who have not yet learned either device.

Left unanswered and holding potential for further research is the question of what the further progression of learning would look like. The fact that the cane (in comparison to the SSD) already reached a certain saturation in the given task spectrum at the end of the study is indicated by several aspects: looking at the performance of B&C, one can roughly estimate the time and average contacts that blind people need to complete the course with their well-trained aid. At the same time, the measurements of S&C are already quite close to those of B&C at the end of the study, so that it can be assumed that only a few more hours of training would be necessary for the sighted to align with them (within the mentioned limited task spectrum of the experimental setup). The assumption that the learning curve of the SSD is less saturated than that of the cane at the end of the study is supported by sighted subjects stating that the cane required less concentration at the end of the study (Q18, 1.8) than the SSD (Q18, 2.7). Furthermore, they expected less learning progress for the cane with "another 3 h of training" (Q23, 2.2) in contrast to the SSD (Q23, 3.5) and also in contrast to the learning progress they already had with the cane during the study (Q22, 3.3). Therefore, a few exciting research questions are whether the learning progress of the SDD would continue in a similar way, at which threshold value it could come to rest and whether this value would eventually be equal to or better than that of the cane. Note that the training time of 2 h in this study is far below that of many other publications on SS; one often-cited study e.g., reached 20–40 h of training with most and 150 h with one particular subject [32].

Another aspect that confounds the interpretation of the data is the presence of a correlation between the two independent variables time/log time and contacts. The reason for this is quite simple: faster walking paces lead to a higher number of contacts. A slow walking pace, on the other hand, allows more time for the interpretation of information and, when approaching an obstacle, to come to a halt in time or to correct the path and not collide with the obstacle. The subjects were asked to consider time and contacts as being equivalent. Yet these variables lack any inherent value that would allow comparing a potential contact with loss of time, for example. Personal preference may also play a role in the weighting of the two variables: fear of collisions with objects (possibly overrepresented in sighted people due to unfamiliarity with the task) may lead to slower speeds. At the same time, the motivation to complete the task particularly quickly may lead to faster speeds but higher collision rates. Several subjects reported that towards the end of the study, they felt that they had learned the device to the point where contacts were completely avoidable for them given some focus. This attitude may have led to a bias in the data, which can be observed in the fact that time increased in the sighted group with both aids towards the end of the study while the collisions continued to fall. It seems to be difficult to solve this problem only by changing the formulation of the task and without expressing the concrete value of an obstacle contact in relation to time (e.g., a collision would add 10 s to the time or leads to the exclusion of this trial from the evaluation).

5.2. Usability & Acceptance

While there is an ongoing debate about how to interpret SUS and what value the scoring system has in the first place, the scores of the SSD (50) and the cane (53) are comparably low. Therefore they can be interpreted as being "OK" only and, in the context of all the systems tested with this score, they tend to be in the 20% or even 10% percentile [69]. It should be noted, however, that the score is rarely used to evaluate assistive devices at all, which may partly explain a generally poorer performance in those. The score is, however, suitable to "compare two versions of an application" [69]: the presented results therefore indicate that usability in the two tested aids does not fundamentally differ in the somewhat small experimental group. This equivalence can also be assessed from other questionnaire items with most having very few deviations:

Looking at the Likert scale averages of the entire study population (blind & sighted), biggest deviations ($\Delta \geq 0.5$) can be observed in the expected recognisability of a "visual impairment because of the aid" that is stated to be much lower (Q15, $\Delta = 2.0$) with the SSD (Q15, 1.8) than with the cane (Q15, 3.8). Just as in the sighted group, the average of both groups stated that the cane was easier to use (Q3, $\Delta = 0.8$) and required less concentration at the beginning of the study (Q17, $\Delta = 0.8$), whereas this difference becomes negligible by the end of the study (Q18, $\Delta = 0.3$). Last but not least, the two aids differed in Q11 ("I had a lot of fun using the aid"), in which the SSD scored $\Delta = 0.7$ points better.

Exemplary statements have already been presented in the results section, summarised and classified into topics. They support the theses that the Unfolding Space Glove achieves its goal of being easier and quicker to learn than many other SSDs while providing users with a positive user experience. However, the sample size and the survey methods are not sufficient for more in-depth analyses. The presentation should rather serve the purpose of completeness and provide insights into how the learning process was perceived by the test persons in the course of the study.

5.3. Distal Attribution and Cognitive Processing

The phenomenon of distal attribution (sometimes externalisation of the stimulus) in simplified terms describes when users of an SSD report to no longer consciously perceive the stimulus at the application site on the body (here e.g., the hand), but instead refer to the perceived objects in space (distal/outside the body). This can also be observed, for example, when sighted people describe visual stimuli and do not describe the perception of stimuli on their retina, but instead the things they *see* at their position in space. Distal attribution was first mentioned in early publications of Paul Bach-y-Rita [31,33] in which participants received 20–40 h of training (one individual even 150 h) with a TVSS and has also been described in other publications e.g., about AVSS devices [40]. Ever since it has been discussed in multifaceted and sometimes even philosophical discourses and has been topic of many experimental investigations [70–73]. As already described in the results section, the statements do not indicate the existence of this specific attribution. However, they do show a high degree of spatio-motor coupling of the stimuli and suggest the emergence of distal-like localisation patterns. The wearing time of only 2 h, however, was comparatively short and studies with longer wearing times would be of great interest on this topic.

5.4. Compliance with the Criteria Set

In the introduction to this paper, 14 criteria were defined that are important for a successful development of an SSD. See also Appendix A for a description of those. Chebat et al. [55], who collected most of them, originally did so to show problems of known SSD proposals. In the design and development process of the Unfolding Space Glove these criteria did play a crucial role from the very start. Now, with the findings of this study in mind, it is time to examine to what degree it can meet the list of criteria by classifying six of its key aspects (with numbers of the respective criteria in parentheses):

- **Open Source and Open Access.** The research, the material, the code and all blueprints of the device, are open source and open access. In the long run, this can lead to lower costs (1) of the SSD and a higher dissemination (4), as development expenses are already eliminated; the device can theoretically even be reproduced by the users themselves.
- **Low Complexity of Information.** The complexity and thus the resolution of the information provided by an SSD is on a low level in order to offer low entry barriers when *learning* (1) the SSD. The *user experience and enjoyment* (13) is not affected much by the *cognitive load* (5). This requirement is of course in contrast to the problems of low *resolution* (9).
- **Usage of 3D Input.** The use of *spatial depth* (7) images from a 3D camera inherently provides suitable information for locomotion, reduces the *cognitive load* (5) of extracting them from conventional two-dimensional images and is independent of lighting situations and bad *contrast* (8).
- **Using Responsive Vibratory Actuators as Output.** Vibration patterns can be felt (in different qualities) all over the body, are non-invasive, non-critical and have no medically proven harmful effect. Linear resonance actuators (LRA), offer low *latency* (3), are durable, do not heat up too much and are still *cost* (10) effective, although they require special integrated circuits to drive them.
- **Positioning at the Back of the Hand.** The site of stimulation on the back of the hand, although disadvantaged by other factors such as total available surface or density of vibrotactile receptors [74], proved to be a suitable site of stimulation with regard to several aspects: a fairly natural posture of the hand when using the device enables a discrete body posture, does not interfere with the overall *aesthetical appearance* (14) and *preserves sensory and motor habits* (12). The *orientation of the Sensor* (6) on the back of the hand is hoped to be quite accurate as we can use our hands for detailed motor actions and have a high proprioceptive precision in our elbow and shoulder [75]. Last but not least, the hand has a *high motor potential* (11) (rotation and movement in three axes), facilitating the sensorimotor coupling process.
- **Thorough Product & Interaction Design.** A good design does not only consist of the visible shell. Functionality, interaction design and product design must be considered holistically and profoundly, and in the end they pay off on many aspects apart from

the *aesthetic appearance* (14) itself, such as almost all of the key aspects discussed in this section and on *user experience and joy of use* (13).

6. Conclusions

The Unfolding Space Glove, a novel wearable haptic spatio-visual sensory substitution system, has been presented in this paper. The glove transforms three-dimensional depth images from a time of flight camera into vibrotactile stimuli on the back of the hand. Blind users can thus haptically explore the depth of the space surrounding them and obstacles contained therein by moving their hand. The device, in its somewhat limited functional scope, can already be used and tested without professional support and without the need of external hardware or specific premises. It already is highly portable and offers a continuous and very immediate feedback, while its design is unobtrusive and discreet.

In a study with eight blind and six sighted (but blindfolded) subjects, the device was tested and evaluated in obstacle courses. It could be shown that all subjects were able to learn the device and successfully complete the parcours presented to them. Handling has low entry barriers and can be learned almost intuitively in a few minutes, with the learning progress between blind and sighted subjects being fairly comparable. However, at the end of the study and after about 2 h of wearing the device, the sighted subjects were significantly slower (by about 54%) in solving the courses with the glove compared to the white long cane they had worn and trained for the same amount of time.

The device meets many basic requirements that a novel SSD has to fulfil in order to be accepted by the target group. This is also reflected in the fact that the participants reported a level of user satisfaction and usability that is—despite its different functions and complexity—quite comparable to that of the white long cane.

The results in the proposed experimental set-up are promising and confirm that depth information presented to the tactile system can be cognitively processed and used to strategically solve navigation tasks. It remains open how much improvement could be achieved in another two or more hours of training with the Unfolding Space Glove. On the other hand, the results are of limited applicability to real-world navigation for blind people: too many basic requirements for a navigation aid system (e.g., detection of ground level objects) are not yet included in the functional spectrum of the device and would have to be implemented and tested in further research.

Supplementary Materials: The following supporting information can be downloaded at: https://www.mdpi.com/article/10.3390/s22051859/s1, S1: GitHub hardware release 0.2.1; S2: GitHub software release 0.2.2; S3: Study data; S4: Exemplary timetable; S5: Exemplary videos (low resolution); S6: GitHub monitor release 1.4. Code, hardware, documentation and building instructions are also available in the public repositories https://github.com/jakobkilian/unfolding-space (accessed on 26 January 2022). Consider Release v0.2.1 for the stable version used in this study and consider more recent commits in which the content has been revised for better accessibility. High resolution video clips of subjects completing trials can also be found at: https://vimeo.com/channels/unfoldingspace (accessed on 26 January 2022). For more information and updates on the project please also see https://www.unfoldingspace.org (accessed on 26 January 2022). All content of the project, including this paper, is licensed under the Creative Commons Attribution (CC-BY-4.0) licence (https://creativecommons.org/licenses/by/4.0/ (accessed on 26 January 2022)). The source code itself is under the MIT licence. Please refer to the LICENSE file in the root directory of the Github repository for detailed information.

Author Contributions: Conceptualization, J.K.; methodology, J.K. and A.N.; software, J.K.; validation, S.W. and L.S.; formal analysis, J.K. and A.N.; investigation, J.K.; resources, J.K. and L.S.; data curation, J.K.; writing—original draft preparation, J.K.; writing—review and editing, A.N., S.W. and L.S.; visualization, J.K.; supervision, A.N., S.W. and L.S.; project administration, J.K.; funding acquisition, S.W. All authors have read and agreed to the published version of the manuscript.

Funding: Funding to conduct the study was received from University of Tübingen (ZUK 63) as part of the German Excellence initiative from the Federal Ministry of Education and Research–Germany (BMBF). This work was done in an industry-on-campus-cooperation between the University of

Tübingen and Carl Zeiss Vision International GmbH. The authors received no specific funding for this work. The funder did not have any role in the study design, data collection and analysis, decision to publish, or preparation of the manuscript. In addition, the prototype construction was funded—also by the BMBF—as part of the Kickstart@TH Cologne project of the StartUpLab@TH Cologne programme ("StartUpLab@TH Cologne", funding reference 13FH015SU8). We furthermore acknowledge support by Open Access Publishing Fund of University of Tübingen.

Institutional Review Board Statement: The study was conducted according to the guidelines of the Declaration of Helsinki, and approved by the Ethics Committee of the faculty of Medicine at the University Hospital of the Eberhard-Karls-University Tübingen (project number 248/2021BO2, approved on 10 May 2021).

Informed Consent Statement: Written informed consent has been obtained from the subjects to publish this paper. Individuals shown in photographs in this paper have explicitly consented to the publication of the photographs by signing an informed consent form.

Data Availability Statement: The data presented in this study are available in Supplementary Materials S3 at: https://www.mdpi.com/article/10.3390/s22051859/s1.

Acknowledgments: The authors would like to thank Kjell Wistoff for his active support in setting up, dismantling and rebuilding the study room, organising the documents and documenting the study photographically; Trainer Regina Beschta for a free introductory O&M course and the loan of the study long cane; Tim Becker and Matthias Krauß from Press Every Key for their open ear when giving advice on software and hardware; The Köln International School of Design (TH Köln) and the responsible parties for making the premises available over this long period of time; pmdtechnologies ag for providing a Pico Flexx camera; Munitec GmbH, for providing glove samples; Connor Shafran for the proofreading of this manuscript; All those who provided guidance in the development of the prototype over the past years and now in the implementation and evaluation of the study. And last but not least to the participants for volunteering for this purpose.

Conflicts of Interest: We declare that Siegfried Wahl is scientist at the University of Tübingen and employee of Carl Zeiss Vision International GmbH, as detailed in the affiliations. There were no conflict of interest regarding this study. The funders had no role in the design of the study; in the collection, analyses, or interpretation of data; in the writing of the manuscript, or in the decision to publish the results.

Abbreviations

The following abbreviations are used in this manuscript:

AIC	Akaike Information Criterion
AVSS	Auditory-Vision Substitution Systems
B&C	blind & cane
B&S	blind & SSD
CI	confidence interval
ERM	eccentric rotating mass
ETA	electronic travel aids
LMM	linear mixed model
LRA	linear resonant actuator
LS	least square (means)
MLE	Maximum Likelihood Estimate
H0	null hypothesis
H1...n	hypothesis 1 to n (numbered)
O&M	orientation & mobility (training)
PS	practice sessions
Q1...n	question 1 to n (numbered)
S&C	sighted & cane
S&S	sighted & SSD
SMC	Sensorimotor Contingencies
SS	Sensory Substitution

SSD	Sensory Substitution Device
SUS	System Usability Score
ToF	Time of Flight
TS	trial session
TVSS	Tactile-Vision Substitution Systems
VIP	visually impaired people

Appendix A. Design Requirements for New SSDs

The following 14 aspects are crucial for the development and evaluation of new SSDs. Points 1 to 10 originate from Chebat et al. [55] while points 11–14 were added by the authors.

1. **Learning:** since mastering an SSD often requires many hours of training, many prospective users are discouraged. In some cases, the motor functions learned by Visually Impaired People (VIPs) to orient themselves are contradictory to functions of SSDs. For example, blind people keep their heads straight in Orientation & Mobility (O&M) training, while using an SSD with a head-mounted camera requires turning the head to get the best scan of the scene. The resulting conflict and fear of losing established and functioning systems through training with SSDs therefore represents a major obstacle.
2. **Training:** most SSD vendors offer no or too little training material for end users. There is also no standardised test procedure that would allow comparisons between systems. A standardised obstacle course proposed by Nau et al. [63] to test low vision or artificial vision is a promising approach, but has, to our knowledge, not yet reached widespread use.
3. **Latency:** some systems, especially AVSS, suffer from high latency between changes in the input image (e.g., by shifting the camera's perspective) and the generated sensory feedback. The resulting low immediacy between motor action by the user and his/her sensory perception, hampers the coupling process with the substituted modality.
4. **Dissemination:** information on available SSDs is simply not very widespread yet. Scientific publications on the topic are often difficult to obtain or not accessible in a barrier-free way for VIPs.
5. **Cognitive load:** with many systems, especially those with high resolution or bandwidth and many output actuators, the interpretation of the stimuli requires a high level of attention and concentration, which is then lacking for other areas of processing, orientation or navigation.
6. **Orientation of the Sensor:** VIPs often find it difficult to determine the real position of objects in the room based on the images perceived with the help of the actuators. This is because the assignment between input and output is not always clear, or it is not apparent which part of the scene they are looking at.
7. **Spatial depth:** for many locomotion navigation tasks, the most important aspect is spatial depth. Extracting this feature from the greyscale image of a complex scene is time-consuming and only possible with good contrast and illumination. If the 3D information is provided directly, there is—in theory—no need for this extraction.
8. **Contrast:** in traditional TVSS systems using conventional passive colour or B/W cameras, illumination and contrast of the scene matter. Using depth information from an actively illuminating 3D cameras (described later on) eliminates this.
9. **Resolution:** by down sampling the resolution of the visual information to fit the physiological limits of the stimulated modality, a lot of acuity and therefore crucial information can be lost. A zoom function is one way to encounter this loss.
10. **Costs:** as the development of SSDs can take many years of research and development, commercially available devices are often expensive, which can deter potential users. Costs can be reduced by using widespread eletronic components instead of specialised parts. Another way to reduce costs is to use existing devices like smartphones and

their features like the camera, the battery and the calculating power. One example for this is the AVSS "the vOICe" [40].

11. **Motor Potential:** as mentioned in the brain plasticity section, a crucial factor for the success of sensorimotor coupling in learning SSDs is movement itself. If the potential for movements with the sensor (hence active motor influence on the sensory input) is low, this enactive process might be hindered. Positioning the sensor or camera on head or trunk, as seen in some examples, influences this potential considerably, as it only has limited rotation and very low potential for movement. If the head is kept straight, as commonly practised in O&M trainings, this potential might be reduced even more.
12. **Preservation of Sensory and Motor Habits:** there is a wide range of possible restrictions of everyday sensory and motor habits that body-worn devices can cause. The overall weight of the device, its form factor and positioning on the body need to be carefully considered and tested, taking into account those habits and needs of VIPs. Blocking the auditory sensory channel by using headphones can, for example, hinder danger-recognition, being addressed and orienting by auditory information. As the hands play a key role in object recognition when feeling objects at close range, their mobility should also be ensured.
13. **User Experience and Joy of Use:** the stimuli should provide a pleasant experience, should not exceed pain thresholds and should not lead to overstraining, irritating or disturbing side effects, even with prolonged use. In order to get used to a new device, not only the purely functional benefit must be convincing. The use of the device should also be enjoyable and trigger a positive user experience. Otherwise, the basic acceptance could be reduced and the learning process could be hindered.
14. **Aesthetic Appearance:** even if functionality is in the foreground, VIPs in particular should not be offered devices that do not meet basic aesthetic standards. A device with a discrete design, which possibly fits stylistically and colour-wise into the outer appearance of the VIP, leads to confidence in it and greater pleasure in using it. At the same time it prevents unnecessary stigmatisation.

Appendix B. Selection of the Input System

ToF cameras have been around since the early 2000s [76], but prices were many times higher than today due to their exclusive use in the industrial segment [77–79]. A few years ago, ToF cameras were increasingly integrated as back-cameras into smartphones to be used for AR applications [80,81]; just recently they (next to similar 3D cameras) became important as a front-camera as well for a more secure biometric face recognition to unlock the device [82].

While there certainly has been work addressing three-dimensional 3D (x, y, depth) input from the environment to aid navigation, the topic is not very prevalent in SS research: some only dealt with sub-areas of a 3D input [83–85] or described ETAs that transmit some kind of simplified depth information, while not being an SSDs in its proper sense [86–90]. In addition, there is work that actually did propose SSDs that use depth data, yet they often had problems implementing or testing the systems in practice because the technology was not yet advanced enough (too slow, heavy and/or expensive) [91–97]. In recent years, however, papers have been published (most of them using sound as an output) that actually proposed and tested applied 3D systems [97–107]. This includes the Sound of Vision device, which today probably ranks among the most advanced and sophisticated, which has also been evaluated in several studies. The project started with auditory feedback only, but now also uses body-worn haptic actuators [101,102,106,108].

Overall, only very few experimented with ToF cameras [89,93] or an array of ToF sensors [103] at all and to the best of our knowledge there is no project that uses a low-cost and comparably fast new generation ToF camera for this, let alone implementing it in a tactile/haptic SSD.

Due to its price, size, form factor, frame rate and resolution the "Pico Flexx" ToF camera development kit ($389 today [79]) was chosen for the prototype (Figure A1).

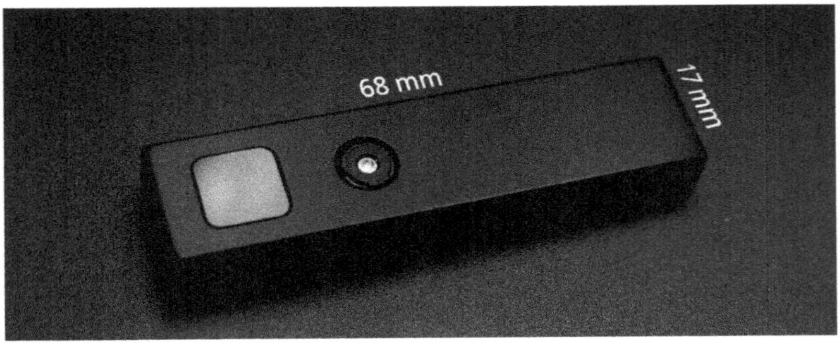

Figure A1. The chosen ToF camera development kit Pico Flexx from pmdtechnologies.

Appendix C. Details on the Setup

Both the *bracelet*—usually used to attach a smartphone for sporting activities—and the 10 Ah *power bank* are commercially available components. The attachment interface was glued to the power bank to be compatible with the bracelet. In the current configuration, the battery lasts about 8 h in ordinary use, with the majority being spent on the computing unit, which has potential to be even more power efficient.

The glove consists of two layers of fabric (modified commercially available gloves, mainly made of polyamide), between which the motors are glued and the cables are sewn. Each motor is covered with a protective sleeve made of heat-shrink tubing to prevent damage to the soldered joints exposed to heavy stress.

Finally there is the *Unfolding Space Carrier Board* attached to the outside using velcro (removable and exchangeable): a printed circuit board specifically made for the project containing the drivers for controlling the motors, the ToF camera attached via USB 3.0 Micro B connector and elastic band, and finally the computing unit—a Raspberry Compute Module 4—connected via a 100-pin mezzanine connector. The CM4101016 configuration of the Compute Module (16 GB Flash, 1 GB RAM, Wifi) that currently is in operation has very little load on flash and RAM and a cheaper version could also be used.

Appendix D. Design of the Actuator System

Once the input side was set up, a suitable interface had to be found to pass on the processed information to a sensory modality using the predefined medium of vibration.

Conventional eccentric rotating mass (ERM) actuators are the first choice for vibratory output in many projects; they are easy to handle, affordable, but not very responsive (rise time starting at 50 ms, usually even higher), noisy and not very durable (100–600 h life time) due to the wear of parts [109]. A series of self-tests with different ERM motors, in different arrangements and on different parts of the body confirmed this and also revealed that the ERMs quickly become uncomfortably hot in continuous operation.

Linear resonant actuators (LRAs) instead provided a remedy in the following tests: with only 10 ms rise time [110] and a higher lifetime (833 h tested, "thousands of hours" possible) [109,110] these are much better suited for this claim; they furthermore consume less power at the same amount of acceleration (which is important for mobile devices) and can apply a higher maximum acceleration [109,110] while in tests remaining cool enough for direct application to the skin.

To keep the complexity on a low level a 3 × 3 LRA matrix proved to be a good set-up in tests. Figure A2 shows this structure on an early prototype.

Figure A2. The matrix of 3 × 3 LRA vibration motors forming the haptic interface of the prototypes—in this case an early prototype is shown.

At this point, it should be mentioned that there is relatively recent research on the perceived differences between these two actuator types when attached to the skin [111,112], which to some extent contradicts these assumptions and instead suggests ERM motors for haptic interfaces. It is yet to be seen what further research in this area will reveal.

Also note that there is a new generation of brush-less direct current (BLDC) ERM motors available not taken into account in this summary, that outperforms classic ERM motors while being more expensive but less energy efficient [109].

Appendix E. Details on the Algorithm

Each depth image is first checked for reliability and then divided into 3 × 3 tiles, each of which is used to create a histogram. Starting from the beginning of the measuring range (~10 cm), these histograms are scanned for objects at each level of depth (0–255) from near to far. If the number of pixels within a range of five depth steps respectively ~ 4 cm exceeds the threshold, the algorithm saves the current distance value of this tile. If however the number remains below the threshold, it is assumed that there is no object within this image tile at this depth level or that there is only image noise. In this case, the algorithm increments the depth step by one and performs the threshold comparison again until it finally arrives at the furthest depth step. The nine values of the resulting 3 × 3 vibration pattern are finally passed on to the vibratory actuators as the amplitude. Each motor thus represents the object that is closest to the camera within the corresponding tile and hence to the hand of the person using the camera.

In Table A1 one can find a summary of the translated modalities. For better illustration, the three-dimensional extension of the field of view of a ToF camera is described as a frustum Figure A3.

Table A1. The translation principle of the algorithm.

Visual Information	Translation into Haptic Stimuli
x-axis of the frustrum (horizontal extension)	x-axis of the motor matrix in 3 levels
y-axis of the frustrum (vertical extension)	y-axis of the motor matrix in 3 levels
z-axis of the frustrum (distance to the camera)	amplitude/vibration strength

To be as platform-independent as possible, a monitoring tool was developed in the Unity 3D game development environment. It receives data from the glove via udp protocol, as long as they are connected to the same (Wifi) network, and displays them visually. This includes the depth image and motor values in real time as well as various technical data such as processing speed, temperature of the Raspberry Pi core and others. It can also be used to control the glove, switch it off temporarily, or test the motors individually.

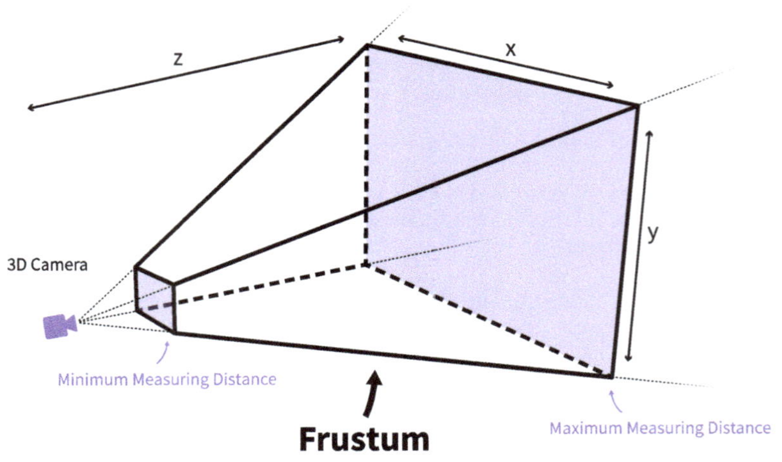

Figure A3. The Volume of the Field of View of a 3D Camera Described as a Frustum.

The files for the Raspberry Pi Code (C++), the monitoring app (Unity 3D, C#) and respective documentation are open source and available on https://github.com/jakobkilian/unfolding-space (accessed on 26 January 2022). Consider Release v0.2.1 for the stable version used in this study and consider more recent commits in which the content has been revised for better accessibility.

Appendix F. Data on Subjects

The following table shows a summary of the subject data from the Supplementary Materials S3. Beyond the information given, none of the subjects used smart canes, none used a guide dog and none had experience with another SSD.

Table A2. Summary of the subject data. Column name abbreviations: ID = subject identifier; VI Level = level of visual impairment according to ICD-10 H54.9 definition; Cane Exp. = years of experience with the cane; O&M = Orientation & Mobility training completed; VA = measured decimal visual acuity (NLP = no light perception, LP = light perception); D. Age = age at diagnosis of full blindness.

ID	Age	Gender	VI Level	Cane Exp.	O&M	VA	D. Age	Description
v	25	m	4 blind	>10 years	yes	LP	1	Cone-rod dystrophy
o	30	f	4 blind	>10 years	yes	NLP	1	Retinal detachment due to premature birth; treatment with oxygen
z	34	f	4 blind	>10 years	yes	LP	1	Retinal detachment
j	37	m	3 blind	1–10 years	yes	0.18	10	Retinitis pigmentosa; cataract (treated); macular edema; visual field 10 degree; visual acuity about 0.2; 10° field of view (IDC Level 3)
h	51	m	5 blind	>10 years	yes	NLP	1	Blindness due to retinal tumour; glass eyes
c	56	m	4 blind	1–10 years	yes	LP	53	Premature birth; incubator; retinopathy affected the eyes; first blindness in the right eye in the 6th year; left eye still VA of 0.1–0.15 for a long time; Blindness in both eyes for 2 years.
y	64	m	5 blind	>10 years	yes	NLP	4	Blindness due to a vaccination
f	65	f	5 blind	>10 years	yes	NLP	0	Birth-blind; toxoplasmosis in pregnancy of mother
u	26	o	0 no VI	-	no	1.25	-	
t	29	f	0 no VI	-	no	1.25	-	
i	32	m	0 no VI	-	no	1.25	-	
e	45	f	0 no VI	-	no	1.25	-	
q	64	m	0 no VI	-	no	1.25	-	
b	72	f	0 no VI	-	no	0.37	-	

Appendix G. Testing Assumptions for Parametric Tests

The model was fitted according to the top to down procedure by Zuur et al. [113] starting from the full model below including all variables that could be of reasonable relevance and piecewise removing those found to be non-significant.

$$(logTime \sim Group * Aid * TS + Order + (TS|Group : Subject) + (1|Layout))$$

Before the model has been reduced, the following assumptions (Figures A4–A6) have been tested in order to be able to use a linear mixed effects model in the first place and later apply parametric tests to it [67]:

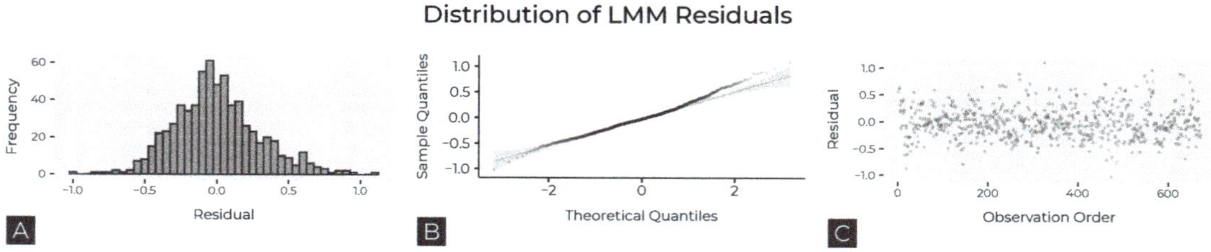

Figure A4. Normal distribution of residuals shown in (**A**) histogram and (**B**) QQ-Plot. Additionally scatterplot in (**C**) shows that there is no major correlation between the residuals.

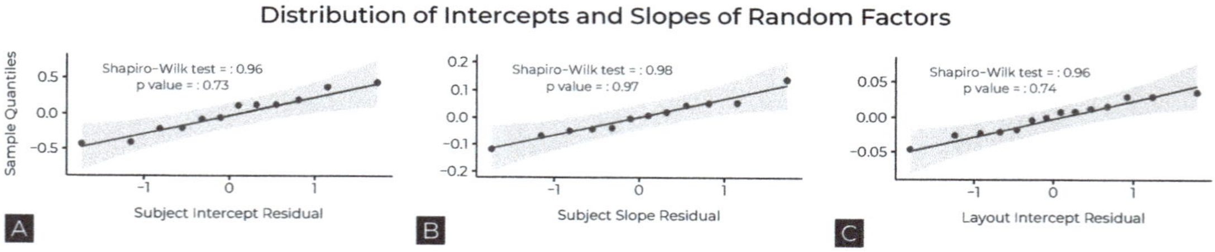

Figure A5. QQ-Plots showing the distribution of the residuals of random effects: (**A**) intercept of subjects, (**B**) slope of subjects (on TS), (**C**) intercept of layout.

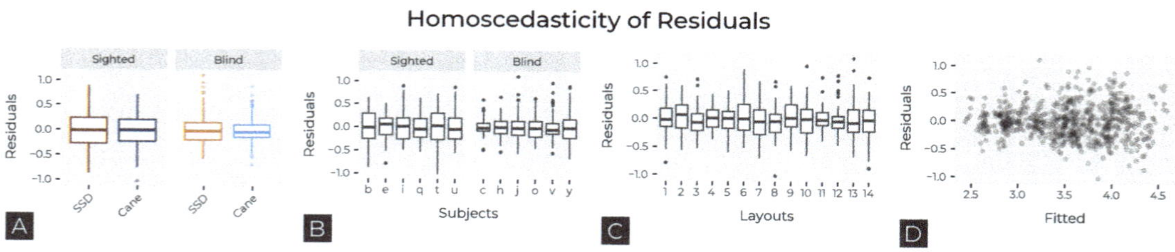

Figure A6. Graphical representations of variance between different variables to show homoscedasticity/homogeneity of variance: (**A**) group and aid, (**B**) subjects and (**C**) layouts. Furthermore the residuals vs. fittet scatterplot in (**D**) allows to check for non-linearity, unequal error variances and outliers. Even though the variances of the error terms seem to be slightly left-slanted, the data can be seen as reasonably homoscedastic and homogeneous.

With these assumptions met, all variables were next tested individually for their effect size on the model [113]. The Akaike Information Criterion (AIC), which is included in Maximum Likelihood Estimates (MLE), was used to identify models that exclude certain variables and therefore show a better fit. Those candidates then have been checked in direct comparisons with a corresponding null model using the Chi-Square p-value. First the order effect (which aid has been tested first in a TS) could be excluded from the full model

($p = 0.886$). Layout has been found to be non-significant as well ($p = 0.144$), however it remains in the model because of reasonable concern about difference in difficulty. Subjects being hierarchically nested within group (due to the unique assignment to one of the two groups) has a significant effect on the model ($p < 0.001$), so do group itself ($p < 0.001$), aid ($p < 0.001$) and TS ($p < 0.001$).

Lastly, several interactions can be observed (Figure A7) in the data (e.g., the blind group already having experience with the cane). The interaction between all three fixed factors (group, aid and TS) has to be included ($p < 0.001$) as well as between TS and aid ($p < 0.001$), between group and aid ($p < 0.001$) and also between aid and TS ($p = 0.007$).

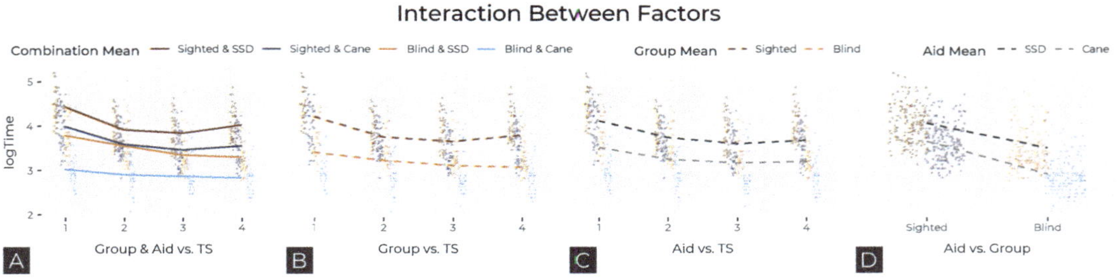

Figure A7. Interaction plot for several variable pairs show interactions in (**A**) group & aid vs. TS, (**B**) in group vs. aid, (**C**) a small interaction in aid vs. TS as well as one in (**D**) aid vs. group.

With order being dropped and the other variables as well as the interactions being kept, the final model was this:

$$(logTime \sim Group * Aid * TS + (TS|Group : Subject) + (1|Layout))$$

Appendix H. Qualitative Statements

In the following appendix, the statements made by the subjects (verbally or in the final questionnaire) are listed in the form of tables, classified by general topics on which the statements are based. All these topics are referred to in the results section as well.

Table A3. Abbreviations used in this table: ID = statement identifier; Subj = subjects; TS = Trial Session.

ID	Statements on Topic 1: Cognitive Processing of the Stimuli	Subj	TS
1	"The glove requires more concentration."	b	1
2	"I'm more occupied with the glove."	i	1
3	"The glove needs a training phase."	q	1
4	"With the glove I need even more time. The cane is easy to handle, but the glove takes longer."	u	1
5	"You could not yet talk, think or do anything else while using the SSD."	c	3
6	"Talking at the same time [when using the SSD] would still be too exhausting."	h	4
7	"You experience the size and distance of objects a bit like touching them. With the cane, on the contrary, I don't imagine myself touching the object."	h	4
8	"I could imagine that with more training it feels like a fabric that gets thicker and thicker the deeper you go. Then you just take the path of least resistance. I even made involuntary movements with my hand at the end of the study."	v	4
9	"Sometimes I just reacted too late [when colliding with an obstacle] and I asked myself: yes, he [the SSD] did warn me, but why didn't I react? With experience, it should become easier."	o	4

Table A4. Abbreviations used in this table: ID = statement identifier; Subj = subjects; TS = Trial Session.

ID	Statements on Topic 2: Perception of Space and Materiality	Subj	TS
1	"Structures are now being formed with the glove: you walk towards something, vibration becomes stronger, an imaginary wall is created. With the cane, on the other hand, it's really there. [But how you can walk around it] can only be determined by three more strokes [with the cane]. With the glove, on the other hand, you directly have the whole picture. If you were to draw it, with the stick it would be points, with the glove it would be lines (with distance). So by structure I mean material or resistance."	v	2
2	"[With the SSD I had] more the feeling of actually seeing. Feeling for distances and gaps was better. More detailed and differentiated perception of the objects."	j	3
3	"[With the SSD] I was able to estimate well how far away the objects were. Partially, a spatial perception of the space in front of me was also possible. [With the SSD] I was able to perceive only what was directly in front of me, not preemptively."	j	4
4	"With the glove I can just look at what's back there on the right."	e	3
5	"Almost »spatial« vision is possible."	e	4
6	"I imagine the object and run my hand along it with the SSD to feel its corners, edges, shape and to know where it ends. [With the glove] I imagine less the object itself, more that there is an obstacle."	t	4

Table A5. Abbreviations used in this table: ID = statement identifier; Subj = subjects; TS = Trial Session.

ID	Statements on Topic 3: Wayfinding Processes	Subj	TS
1	"I use the glove strategically: I can look further to the left or right and use it to find my way."	v	1
2	"With the glove, you can orientate yourself earlier and make decisions sooner."	i	3
3	"It is more a matter of seeking and finding the obstacles with the cane."	y	4
4	"With the glove you can anticipate better. With the cane you are dependent on the collision of the obstacle with the cane (*bang bang*) and then have to run somewhere else without knowing what is coming. With the glove you run directly towards the spot that is free: you first check the surroundings and then think about where you are going and don't just start running and see what is coming."	j	1
5	"With the cane, you only notice it when you hit it. With the glove you can react before that."	o	4
6	"The cane is more like hand-to-hand combat."	c	1
7	"The cane makes you feel clumsy. It's annoying that you make a loud noise with it"	u	1

Table A6. Abbreviations used in this table: ID = statement identifier; Subj = subjects; TS = Trial Session.

ID	Statements on Topic 4: Enjoyment of Use	Subj	TS
1	"With the stick it's not so much fun because you can already master it safely."	t	3
2	[Talking about the glove:] "That was fun!"	i	3
3	"I could imagine running only with the glove in certain situations. The glove is fun. You can really immerse yourself in it."	v	4

Table A7. Abbreviations used in this table: ID = statement identifier; Subj = subjects; TS = Trial Session.

ID	Statements on Topic 5: Feeling Safe and Comfortable	Subj	TS
1	"With the cane I'm not nervous, I walked fast, I'm very confident. In the last two runs with the glove I also felt more confident."	t	3
2	"Cane was always safe, glove got better towards the end and I was less afraid."	u	1
3	"With the cane I already felt comfortable the last time. Can run faster with stick without anything happening. More unsafe with glove."	i	3

Table A8. Abbreviations used in this table: ID = statement identifier; Subj = subjects; TS = Trial Session.

ID	Statements on Topic 6: Distal Attribution	Subj	TS
1	"I imagine the object and run my hand along it with the SSD to feel its corners, edges, shape and to know where it ends."	t	4
2	"Got more of a feeling of really seeing [With the SSD]."	j	4
3	"[With the SSD] I was able to estimate well how far away the objects were. Partially, a spatial idea of the space in front of me was also possible."	j	4
4	"This time I became aware of the change of space through my movement"	i	3
5	"Structures are forming [. . .]. You start to decode the information and build a space."	v	2
6	"[With the SSD] you instinctively get the feeling »there's something ahead« and you want to get out of the way."	t	4

References

1. Proulx, M.J.; Gwinnutt, J.; Dell'Erba, S.; Levy-Tzedek, S.; de Sousa, A.A.; Brown, D.J. Other Ways of Seeing: From Behavior to Neural Mechanisms in the Online "Visual" Control of Action with Sensory Substitution. *Restor. Neurol. Neurosci.* **2016**, *34*, 29–44. [CrossRef]
2. Jacobson, H. The Informational Capacity of the Human Eye. *Science* **1951**, *113*, 292–293. . [CrossRef]
3. Montarzino, A.; Robertson, B.; Aspinall, P.; Ambrecht, A.; Findlay, C.; Hine, J.; Dhillon, B. The Impact of Mobility and Public Transport on the Independence of Visually Impaired People. *Vis. Impair. Res.* **2007**, *9*, 67–82. [CrossRef]
4. Taipale, J.; Mikhailova, A.; Ojamo, M.; Nättinen, J.; Väätäinen, S.; Gissler, M.; Koskinen, S.; Rissanen, H.; Sainio, P.; Uusitalo, H. Low Vision Status and Declining Vision Decrease Health-Related Quality of Life: Results from a Nationwide 11-Year Follow-up Study. *Qual. Life Res.* **2019**, *28*, 3225–3236. [CrossRef]
5. Brown, R.L.; Barrett, A.E. Visual Impairment and Quality of Life Among Older Adults: An Examination of Explanations for the Relationship. *J. Gerontol. Ser. B* **2011**, *66*, 364–373. [CrossRef]
6. Hersh, M.A.; Johnson, M.A. Disability and Assistive Technology Systems. In *Assistive Technology for Visually Impaired and Blind People*; Hersh, M.A., Johnson, M.A., Eds.; Springer: London, UK, 2008; pp. 1–50. [CrossRef]
7. Hersh, M.A.; Johnson, M.A. *Assistive Technology for Visually Impaired and Blind People*; Springer: London, UK, 2008.
8. Pissaloux, E.; Velazquez, R. (Eds.) *Mobility of Visually Impaired People: Fundamentals and ICT Assistive Technologies*; Springer International Publishing: Cham, Switzerland, 2018.
9. Montello, D.R.; Sas, C. Human Factors of Wayfinding in Navigation. In *International Encyclopedia of Ergonomics and Human Factors*; CRC Press/Taylor & Francis, Ltd.: London, UK, 2006; pp. 2003–2008.
10. Goldschmidt, M. Orientation and Mobility Training to People with Visual Impairments. In *Mobility of Visually Impaired People: Fundamentals and ICT Assistive Technologies*; Pissaloux, E., Velazquez, R., Eds.; Springer International Publishing: Cham, Switzerland, 2018; pp. 237–261. [CrossRef]
11. Loomis, J.M.; Klatzky, R.L.; Golledge, R.G.; Cicinelli, J.G.; Pellegrino, J.W.; Fry, P.A. Nonvisual Navigation by Blind and Sighted: Assessment of Path Integration Ability. *J. Exp. Psychol. Gen.* **1993**, *122*, 73–91. [CrossRef]
12. Kusnyerik, A.; Resch, M.; Kiss, H.; Nemeth, J. Vision Restoration with Implants. In *Mobility of Visually Impaired People: Fundamentals and ICT Assistive Technologies*; Pissaloux, E., Velazquez, R., Eds.; Springer International Publishing: Cham, Switzerland, 2018; pp. 617–630. [CrossRef]
13. Edwards, T.L.; Cottriall, C.L.; Xue, K.; Simunovic, M.P.; Ramsden, J.D.; Zrenner, E.; MacLaren, R.E. Assessment of the Electronic Retinal Implant Alpha AMS in Restoring Vision to Blind Patients with End-Stage Retinitis Pigmentosa. *Ophthalmology* **2018**, *125*, 432–443. [CrossRef]
14. Schinazi, V.R.; Thrash, T.; Chebat, D.R. Spatial Navigation by Congenitally Blind Individuals. *Wiley Interdiscip. Rev. Cogn. Sci.* **2016**, *7*, 37–58. [CrossRef]
15. Giudice, N.A. Navigating without Vision: Principles of Blind Spatial Cognition. In *Handbook of Behavioral and Cognitive Geography*; Edward Elgar Publishing: Northampton, MA, USA, 2018; pp. 260–288. [CrossRef]
16. Roentgen, U.R.; Gelderblom, G.J.; Soede, M.; de Witte, L.P. Inventory of Electronic Mobility Aids for Persons with Visual Impairments: A Literature Review. *J. Vis. Impair. Blind.* **2008**, *102*, 702–724. [CrossRef]
17. Gold, D.; Simson, H. Identifying the Needs of People in Canada Who Are Blind or Visually Impaired: Preliminary Results of a Nation-Wide Study. *Int. Congr. Ser.* **2005**, *1282*, 139–142. [CrossRef]
18. Leonard, R.M.; Gordon, A.R. *Statistics on Vision Impairment: A Resource Manual*; Lighthouse International: New York, NY, USA, 2001.
19. Manduchi, R. Mobility-Related Accidents Experienced by People with Visual Impairment. *Insight Res. Pract. Vis. Impair. Blind.* **2011**, *4*, 44–54.
20. Kim, S.Y.; Cho, K. Usability and Design Guidelines of Smart Canes for Users with Visual Impairments. *Int. J. Des.* **2013**, *7*, 99–110.
21. Farmer, L.W. Mobility Devices. In *Foundations of Orientation and Mobility*; Welsh, R.L., Blasch, B.B., Eds.; American Foundation for the Blind: New York, NY, USA, 1980; pp. 357–412.
22. Buchs, G.; Simon, N.; Maidenbaum, S.; Amedi, A. Waist-up Protection for Blind Individuals Using the EyeCane as a Primary and Secondary Mobility Aid. *Restor. Neurol. Neurosci.* **2017**, *35*, 225–235. [CrossRef]
23. dos Santos, A.D.P.; Medola, F.O.; Cinelli, M.J.; Garcia Ramirez, A.R.; Sandnes, F.E. Are Electronic White Canes Better than Traditional Canes? A Comparative Study with Blind and Blindfolded Participants. *Univ. Access Inf. Soc.* **2021**, *20*, 93–103. [CrossRef]
24. Dakopoulos, D.; Bourbakis, N.G. Wearable Obstacle Avoidance Electronic Travel Aids for Blind: A Survey. *IEEE Trans. Syst. Man Cybern. Part C Appl. Rev.* **2010**, *40*, 25–35. [CrossRef]
25. Robert H.W. Dog Guides. In *Foundations of Orientation and Mobility*; Welsh, R.L., Blasch, B.B., Eds.; American Foundation for the Blind: New York, NY, USA, 1980; pp. 565–580.
26. Whitmarsh, L. The Benefits of Guide Dog Ownership. *Vis. Impair. Res.* **2005**, *7*, 27–42. [CrossRef]
27. Wiggett-Barnard, C.; Steel, H. The Experience of Owning a Guide Dog. *Disabil. Rehabil.* **2008**, *30*, 1014–1026. [CrossRef]
28. Knol, B.W.; Roozendaal, C.; van den Bogaard, L.; Bouw, J. The Suitability of Dogs as Guide Dogs for the Blind: Criteria and Testing Procedures. *Vet. Q.* **1988**, *10*, 198–204. [CrossRef]

29. Bourne, R.; Steinmetz, J.D.; Flaxman, S.; Briant, P.S.; Taylor, H.R.; Resnikoff, S.; Casson, R.J.; Abdoli, A.; Abu-Gharbieh, E.; Afshin, A.; et al. Trends in Prevalence of Blindness and Distance and near Vision Impairment over 30 Years: An Analysis for the Global Burden of Disease Study. *Lancet Glob. Health* **2021**, *9*, e130–e143. [CrossRef]
30. Bach-y-Rita, P.; Kercel, S.W. Sensory Substitution and the Human-Machine Interface. *Trends Cogn. Sci. (Regul. Ed.)* **2003**, *7*, 541–546. [CrossRef]
31. Bach-y-Rita, P. *Brain Mechanisms in Sensory Substitution*; Academic Press Inc.: New York, NY, USA, 1972.
32. Bach-y-Rita, P.; Collins, C.C.; Saunders, F.A.; White, B.; Scadden, L. Vision Substitution by Tactile Image Projection. *Nature* **1969**, *221*, 963–964. [CrossRef]
33. White, B.W.; Saunders, F.A.; Scadden, L.; Bach-Y-Rita, P.; Collins, C.C. Seeing with the Skin. *Percept. Psychophys.* **1970**, *7*, 23–27. [CrossRef]
34. Bach-y-Rita, P.; Tyler, M.E.; Kaczmarek, K.A. Seeing with the Brain. *Int. J. Hum.–Comput. Interact.* **2003**, *15*, 285–295. [CrossRef]
35. McDaniel, T.; Krishna, S.; Balasubramanian, V.; Colbry, D.; Panchanathan, S. Using a Haptic Belt to Convey Non-Verbal Communication Cues during Social Interactions to Individuals Who Are Blind. In Proceedings of the 2008 IEEE International Workshop on Haptic Audio Visual Environments and Games, Ottawa, ON, Canada, 18–19 October 2008; pp. 13–18. [CrossRef]
36. Krishna, S.; Bala, S.; McDaniel, T.; McGuire, S.; Panchanathan, S. VibroGlove: An Assistive Technology Aid for Conveying Facial Expressions. In Proceedings of the CHI '10 Extended Abstracts on Human Factors in Computing Systems, Atlanta, GA, USA, 10–15 April 2010; ACM Press: New York, NY, USA, 2010; pp. 3637–3642. [CrossRef]
37. Hanneton, S.; Auvray, M.; Durette, B. The Vibe: A Versatile Vision-to-Audition Sensory Substitution Device. *Appl. Bionics Biomech.* **2010**, *7*, 269–276. [CrossRef]
38. Zelek, J.S.; Bromley, S.; Asmar, D.; Thompson, D. A Haptic Glove as a Tactile-Vision Sensory Substitution for Wayfinding. *J. Vis. Impair. Blind.* **2003**, *97*, 621–632. [CrossRef]
39. Loomis, J.M.; Golledge, R.G.; Klatzky, R.L. Navigation System for the Blind: Auditory Display Modes and Guidance. *Presence Teleoperat. Virtual Environ.* **1998**, *7*, 193–203. [CrossRef]
40. Meijer, P. An Experimental System for Auditory Image Representations. *IEEE Trans. Biomed. Eng.* **1992**, *39*, 112–121. [CrossRef]
41. Sampaio, E.; Maris, S.; Bach-y-Rita, P. Brain Plasticity: 'Visual' Acuity of Blind Persons via the Tongue. *Brain Res.* **2001**, *908*, 204–207. [CrossRef]
42. Haigh, A.; Brown, D.J.; Meijer, P.; Proulx, M.J. How Well Do You See What You Hear? The Acuity of Visual-to-Auditory Sensory Substitution. *Front. Psychol.* **2013**, *4*, 330. [CrossRef]
43. Abboud, S.; Hanassy, S.; Levy-Tzedek, S.; Maidenbaum, S.; Amedi, A. EyeMusic: Introducing a "Visual" Colorful Experience for the Blind Using Auditory Sensory Substitution. *Restor. Neurol. Neurosci.* **2014**, *32*, 247–257. [CrossRef]
44. Lenay, C.; Gapenne, O.; Hanneton, S.; Marque, C.; Genouëlle, C. Chapter 16. Sensory Substitution: Limits and Perspectives. In *Advances in Consciousness Research*; Hatwell, Y., Streri, A., Gentaz, E., Eds.; John Benjamins Publishing Company: Amsterdam, The Netherlands, 2003; Volume 53, pp. 275–292. [CrossRef]
45. Velázquez, R. Wearable Assistive Devices for the Blind. In *Wearable and Autonomous Biomedical Devices and Systems for Smart Environment*; Lecture Notes in Electrical Engineering; Springer: Berlin/Heidelberg, Germany, 2010; pp. 331–349. [CrossRef]
46. Dunai, L.D.; Lengua, I.L.; Tortajada, I.; Simon, F.B. Obstacle Detectors for Visually Impaired People. In Proceedings of the 2014 International Conference on Optimization of Electrical and Electronic Equipment (OPTIM), Bran, Romania, 22–24 May 2014; pp. 809–816. [CrossRef]
47. Radar Glasses. Available online: https://www.instructables.com/Radar-Glasses (accessed on 26 January 2022).
48. Stuff Made Here. See in Complete Darkness with Touch. 2020. Available online: https://www.youtube.com/watch?v=8Au47gnXs0w (accessed on 26 January 2022).
49. HaptiVision. Available online: https://hackaday.io/project/27112-haptivision (accessed on 26 January 2022).
50. Low Cost Tongue Vision. Available online: https://hackaday.io/project/13446-low-cost-tongue-vision (accessed on 26 January 2022).
51. Bach-y-Rita, P. Recovery from Brain Damage. *J. Neurol. Rehabil.* **1992**, *6*, 191–199. [CrossRef]
52. Bach-y-Rita, P. Nonsynaptic Diffusion Neurotransmission and Brain Plasticity. *Neuroscientist* **1996**, *2*, 260–261. [CrossRef]
53. Held, R.; Hein, A. Movement-Produced Stimulation in the Development of Visually Guided Behavior. *J. Comp. Physiol. Psychol.* **1963**, *56*, 872–876. [CrossRef]
54. O'Regan, J.K.; Noë, A. A Sensorimotor Account of Vision and Visual Consciousness. *Behav. Brain Sci.* **2001**, *24*, 939–973. [CrossRef]
55. Chebat, D.R.; Harrar, V.; Kupers, R.; Maidenbaum, S.; Amedi, A.; Ptito, M. Sensory Substitution and the Neural Correlates of Navigation in Blindness. In *Mobility of Visually Impaired People*; Pissaloux, E., Velazquez, R., Eds.; Springer: Cham, Switzerland, 2018; pp. 167–200. [CrossRef]
56. Elli, G.V.; Benetti, S.; Collignon, O. Is There a Future for Sensory Substitution Outside Academic Laboratories? *Multisens. Res.* **2014**, *27*, 271–291. [CrossRef]
57. Bhowmick, A.; Hazarika, S.M. An Insight into Assistive Technology for the Visually Impaired and Blind People: State-of-the-Art and Future Trends. *J. Multimodal User Interfaces* **2017**, *11*, 149–172. [CrossRef]
58. Kilian, J. Sensorische Plastizität—Die Körperliche Konstitution von Kognition Als Erklärungsmodell Sensorischer Substitution Und Augmentation. Bachelor's Thesis, TH Köln—Köln International School of Design, Köln, Germany, 2018. [CrossRef]

59. Kilian, J. Unfolding Space–Haptische Wahrnehmung Räumlicher Bilder Durch Sensorische Substitution. Bachelor's Thesis, TH Köln—Köln International School of Design, Köln, Germany, 2018. [CrossRef]
60. pmdtechnologies ag. Development Kit Brief–CamBoard Pico Flexx. 2017. Available online: https://pmdtec.com/picofamily/wp-content/uploads/2018/03/PMD_DevKit_Brief_CB_pico_flexx_CE_V0218-1.pdf (accessed on 26 January 2022).
61. Li, L. *Time-of-Flight Camera—An Introduction*; Technical White Paper; Texas Instruments: Dallas, TX, USA, 2014.
62. Bach, M. The Freiburg Visual Acuity Test—Automatic Measurement of Visual Acuity. *Optom. Vis. Sci.* **1996**, *73*, 49–53. [CrossRef] [PubMed]
63. Nau, A.C.; Pintar, C.; Fisher, C.; Jeong, J.H.; Jeong, K. A Standardized Obstacle Course for Assessment of Visual Function in Ultra Low Vision and Artificial Vision. *J. Vis. Exp.* **2014**, *84*, e51205. [CrossRef]
64. Due, B.L.; Kupers, R.; Lange, S.B.; Ptito, M. Technology Enhanced Vision in Blind and Visually Impaired Individuals. *Circd Work. Pap. Soc. Interact.* **2017**, *3*, 1–31.
65. Paré, S.; Bleau, M.; Djerourou, I.; Malotaux, V.; Kupers, R.; Ptito, M. Spatial Navigation with Horizontally Spatialized Sounds in Early and Late Blind Individuals. *PLoS ONE* **2021**, *16*, e0247448. [CrossRef]
66. Brooke, J. SUS: A 'Quick and Dirty' Usability Scale. In *Usability Evaluation in Industry*; Jordan, P.W., Thomas, B., Weerdmeester, B.A., McClelland, A.L., Eds.; Taylor and Francis: London, UK, 1996; Volume 189, pp. 189–194.
67. Zuur, A.F.; Ieno, E.N.; Elphick, C.S. A Protocol for Data Exploration to Avoid Common Statistical Problems. *Methods Ecol. Evol.* **2010**, *1*, 3–14. [CrossRef]
68. Walker, E.; Nowacki, A.S. Understanding Equivalence and Noninferiority Testing. *J. Gen. Intern. Med.* **2011**, *26*, 192–196. [CrossRef] [PubMed]
69. Brooke, J. SUS: A Retrospective. *J. Usability Stud.* **2013**, *8*, 29–40.
70. Epstein, W.; Hughes, B.; Schneider, S.; Bach-y-Rita, P. Is There Anything out There? A Study of Distal Attribution in Response to Vibrotactile Stimulation. *Perception* **1986**, *15*, 275–284. [CrossRef]
71. Loomis, J.M. Distal Attribution and Presence. *Presence Teleoperat. Virtual Environ.* **1992**, *1*, 113–119. [CrossRef]
72. Auvray, M.; Hanneton, S.; Lenay, C.; O'regan, K. There Is Something out There: Distal Attribution in Sensory Substitution, Twenty Years Later. *J. Integr. Neurosci.* **2005**, *4*, 505–521. [CrossRef]
73. Siegle, J.H.; Warren, W.H. Distal Attribution and Distance Perception in Sensory Substitution. *Perception* **2010**, *39*, 208–223. [CrossRef] [PubMed]
74. Sato, T.; Okada, Y.; Miyamoto, T.; Fujiyama, R. Distributions of Sensory Spots in the Hand and Two-Point Discrimination Thresholds in the Hand, Face and Mouth in Dental Students. *J. Physiol.-Paris* **1999**, *93*, 245–250. [CrossRef]
75. van Beers, R.J.; Sittig, A.C.; Denier van der Gon, J.J. The Precision of Proprioceptive Position Sense. *Exp. Brain Res.* **1998**, *122*, 367–377. [CrossRef]
76. Oggier, T.; Lehmann, M.; Kaufmann, R.; Schweizer, M.; Richter, M.; Metzler, P.; Lang, G.; Lustenberger, F.; Blanc, N. An All-Solid-State Optical Range Camera for 3D Real-Time Imaging with Sub-Centimeter Depth Resolution (SwissRanger). *Soc. Photogr. Instrum. Eng.* **2004**, *5249*, 534–545. [CrossRef]
77. Kuhlmann, U. 3D-Bilder mit Webcams. 2004. Available online: https://www.heise.de/newsticker/meldung/3D-Bilder-mit-Webcams-121098.html (accessed on 26 January 2022).
78. pmdtechnologies ag. Der 3D-Tiefensensor CamBoard pico flexx ist für vielfältige Anwendungsfälle und Zielgruppen Optimiert. 2017. Available online: https://www.pressebox.de/pressemitteilung/pmdtechnologies-gmbh/Der-3D-Tiefensensor-CamBoard-pico-flexx-ist-fuer-vielfaeltige-Anwendungsfaelle-und-Zielgruppen-optimiert/boxid/867050 (accessed on 26 January 2022).
79. Development Kit Pmd[Vision]® CamBoard Pico Flexx. Available online: https://www.automation24.com/development-kit-pmd-vision-r-camboard-pico-flexx-700-000-094 (accessed on 26 January 2022).
80. pmdtechnologies ag. Sensing the World the Way We Do—The World's First Project Tango Smartphone. 2016. Available online: https://www.pressebox.com/pressrelease/pmdtechnologies-gmbh/Sensing-the-world-the-way-we-do-the-worlds-first-Project-Tango-Smartphone/boxid/800705 (accessed on 26 January 2022).
81. pmdtechnologies ag. Pmdtechnologies Coming Strong. 2017. Available online: https://pmdtec.com/fileadmin/user_upload/PR_20170501_pmdinsideASUSphone.pdf (accessed on 26 January 2022).
82. Das, A.; Galdi, C.; Han, H.; Ramachandra, R.; Dugelay, J.L.; Dantcheva, A. Recent Advances in Biometric Technology for Mobile Devices. In Proceedings of the 2018 IEEE 9th International Conference on Biometrics Theory, Applications and Systems (BTAS), Redondo Beach, CA, USA, 22–25 October 2018; pp. 1–11. [CrossRef]
83. Kawai, Y.; Kobayashi, M.; Minagawa, H.; Miyakawa, M.; Tomita, F. A Support System for Visually Impaired Persons Using Three-Dimensional Virtual Sound. *IEEJ Trans. Electron. Inf. Syst.* **2000**, *120*, 648–655. [CrossRef]
84. Lu, X.; Manduchi, R. Detection and Localization of Curbs and Stairways Using Stereo Vision. In Proceedings of the 2005 IEEE International Conference on Robotics and Automation, Barcelona, Spain, 18–22 April 2005; pp. 4648–4654. [CrossRef]
85. Lee, D.J.; Anderson, J.D.; Archibald, J.K. Hardware Implementation of a Spline-Based Genetic Algorithm for Embedded Stereo Vision Sensor Providing Real-Time Visual Guidance to the Visually Impaired. *EURASIP J. Adv. Signal Process.* **2008**, *2008*, 1–10. [CrossRef]
86. Bourbakis, N. Sensing Surrounding 3-D Space for Navigation of the Blind. *IEEE Eng. Med. Biol. Mag.* **2008**, *27*, 49–55. [CrossRef]

87. Zöllner, M.; Huber, S.; Jetter, H.C.; Reiterer, H. NAVI—A Proof-of-Concept of a Mobile Navigational Aid for Visually Impaired Based on the Microsoft Kinect. In Proceedings of the Human-Computer Interaction—INTERACT 2011, Lisbon, Portugal, 5–9 September 2011; Lecture Notes in Computer Science; Campos, P., Graham, N., Jorge, J., Nunes, N., Palanque, P., Winckler, M., Eds.; Springer: Berlin/Heidelberg, Germany, 2011; pp. 584–587. [CrossRef]
88. Dunai, L.; Peris-Fajarnés, G.; Lluna, E.; Defez, B. Sensory Navigation Device for Blind People. *J. Navig.* **2013**, *66*, 349–362. [CrossRef]
89. Lengua, I.; Dunai, L.; Peris-Fajarnés, G.; Defez, B. Navigation Device for Blind People Based on Time-of-Flight Technology. *Dyna-Colombia* **2013**, *80*, 33–41.
90. Mattoccia, S.; Macrı', P. 3D Glasses as Mobility Aid for Visually Impaired People. In Proceedings of the Computer Vision—ECCV 2014 Workshops, Zurich, Switzerland, 6–12 September 2014; Lecture Notes in Computer Science; Agapito, L., Bronstein, M.M., Rother, C., Eds.; Springer International Publishing: Cham, Switzerland, 2015; pp. 539–554. [CrossRef]
91. Balakrishnan, G.; Sainarayanan, G.; Nagarajan, R.; Yaacob, S. Wearable Real-Time Stereo Vision for the Visually Impaired. *Eng. Lett.* **2007**, *14*, 6–14.
92. Bujacz, M. Representing 3D Scenes through Spatial Audio in an Electronic Travel Aid for the Blind. Ph.D. Thesis, Technical University of Lodz, Faculty of Electrical, Electronic, Computer and Control Engineering, Łódź, Poland, 2010.
93. Callaghan, K.J.O.; Mahony, M.J.O. An Obstacle Segmentation and Classification System for the Visually Impaired. In Proceedings of the 2010 International Symposium on Optomechatronic Technologies, Toronto, ON, Canada, 24–27 October 2010; pp. 1–6. [CrossRef]
94. Akhter, S.; Mirsalahuddin, J.; Marquina, F.; Islam, S.; Sareen, S. A Smartphone-based Haptic Vision Substitution System for the Blind. In Proceedings of the 2011 IEEE 37th Annual Northeast Bioengineering Conference (NEBEC), Troy, NY, USA, 1–3 April 2011; IEEE: Troy, NY, USA, 2011; pp. 1–2. [CrossRef]
95. Cancar, L.; Díaz, A.; Barrientos, A.; Travieso, D.; Jacobs, D.M. Tactile-Sight: A Sensory Substitution Device Based on Distance-Related Vibrotactile Flow. *Int. J. Adv. Robot. Syst.* **2013**, *10*, 272. [CrossRef]
96. Maidenbaum, S.; Chebat, D.R.; Levy-Tzedek, S.; Amedi, A. Depth-To-Audio Sensory Substitution for Increasing the Accessibility of Virtual Environments. In *Universal Access in Human-Computer Interaction. Design and Development Methods for Universal Access*; Lecture Notes in Computer Science; Stephanidis, C., Antona, M., Eds.; Springer International Publishing: Cham, Switzerland, 2014; pp. 398–406. [CrossRef]
97. Stoll, C.; Palluel-Germain, R.; Fristot, V.; Pellerin, D.; Alleysson, D.; Graff, C. Navigating from a Depth Image Converted into Sound. *Appl. Bionics Biomech.* **2015**, *2015*, 543492. [CrossRef]
98. Filipe, V.; Fernandes, F.; Fernandes, H.; Sousa, A.; Paredes, H.; Barroso, J. Blind Navigation Support System Based on Microsoft Kinect. *Procedia Comput. Sci.* **2012**, *14*, 94–101. [CrossRef]
99. Gholamalizadeh, T.; Pourghaemi, H.; Mhaish, A.; Ince, G.; Duff, D.J. Sonification of 3D Object Shape for Sensory Substitution: An Empirical Exploration. In Proceedings of the ACHI 2017, The Tenth International Conference on Advances in Computer-Human Interactions, Nice, France, 19–23 March 2017; pp. 18–24.
100. Morar, A.; Moldoveanu, F.; Petrescu, L.; Moldoveanu, A. Real Time Indoor 3D Pipeline for an Advanced Sensory Substitution Device. In Proceedings of the Image Analysis and Processing—ICIAP 2017, Catania, Italy, 11–15 September 2017; Lecture Notes in Computer Science; Springer: Cham, Switzerland, 2017; pp. 685–695. [CrossRef]
101. Caraiman, S.; Morar, A.; Owczarek, M.; Burlacu, A.; Rzeszotarski, D.; Botezatu, N.; Herghelegiu, P.; Moldoveanu, F.; Strumillo, P.; Moldoveanu, A. Computer Vision for the Visually Impaired: The Sound of Vision System. In Proceedings of the 2017 IEEE International Conference on Computer Vision Workshops (ICCVW), Venice, Italy, 22–29 October 2017; pp. 1480–1489. [CrossRef]
102. Caraiman, S.; Zvoristeanu, O.; Burlacu, A.; Herghelegiu, P. Stereo Vision Based Sensory Substitution for the Visually Impaired. *Sensors* **2019**, *19*, 2771. [CrossRef]
103. Katzschmann, R.K.; Araki, B.; Rus, D. Safe Local Navigation for Visually Impaired Users with a Time-of-Flight and Haptic Feedback Device. *IEEE Trans. Neural Syst. Rehabil. Eng.* **2018**, *26*, 583–593. [CrossRef] [PubMed]
104. Massiceti, D.; Hicks, S.L.; van Rheede, J.J. Stereosonic Vision: Exploring Visual-to-Auditory Sensory Substitution Mappings in an Immersive Virtual Reality Navigation Paradigm. *PLoS ONE* **2018**, *13*, e0199389. [CrossRef] [PubMed]
105. Neugebauer, A.; Rifai, K.; Getzlaff, M.; Wahl, S. Navigation Aid for Blind Persons by Visual-to-Auditory Sensory Substitution: A Pilot Study. *PLoS ONE* **2020**, *15*, e0237344. [CrossRef] [PubMed]
106. Zvorișteanu, O.; Caraiman, S.; Lupu, R.G.; Botezatu, N.A.; Burlacu, A. Sensory Substitution for the Visually Impaired: A Study on the Usability of the Sound of Vision System in Outdoor Environments. *Electronics* **2021**, *10*, 1619. [CrossRef]
107. Hamilton-Fletcher, G.; Alvarez, J.; Obrist, M.; Ward, J. SoundSight: A Mobile Sensory Substitution Device That Sonifies Colour, Distance, and Temperature. *J. Multimodal User Interfaces* **2021**, *16*, 107–123. [CrossRef]
108. Poryzala, P. EEG Measures of Auditory and Tactile Stimulations in Computer Vision Based Sensory Substitution System for Visually Impaired Users. In Proceedings of the 2018 11th International Conference on Human System Interaction (HSI), Gdańsk, Poland, 4–6 July 2018; pp. 200–206. [CrossRef]
109. Precision Microdrives. AB-028: Vibration Motor Comparison Guide. 2021. Available online: https://www.precisionmicrodrives.com/ab-028 (accessed on 26 January 2022).
110. Electronics, J.M. G1040003D—Product Specification. 2019. Available online: https://vibration-motor.com/coin-vibration-motors/coin-linear-resonant-actuators-lra/g1040003d (accessed on 26 January 2022).

111. Seim, C.; Hallam, J.; Raghu, S.; Le, T.A.; Bishop, G.; Starner, T. *Perception in Hand-Worn Haptics: Placement, Simultaneous Stimuli, and Vibration Motor Comparisons*; Technical Report; Georgia Institute of Technology: Atlanta, GA, USA, 2015.
112. Huang, H.; Li, T.; Antfolk, C.; Enz, C.; Justiz, J.; Koch, V.M. Experiment and Investigation of Two Types of Vibrotactile Devices. In Proceedings of the 2016 6th IEEE International Conference on Biomedical Robotics and Biomechatronics (BioRob), Singapore, 26–29 June 2016; pp. 1266–1271. [CrossRef]
113. Zuur, A.F.; Ieno, E.N.; Walker, N.; Saveliev, A.A.; Smith, G.M. *Mixed Effects Models and Extensions in Ecology with R*; Statistics for Biology and Health; Springer: New York, NY, USA, 2009. [CrossRef]

sensors

MDPI

Article

Colorophone 2.0: A Wearable Color Sonification Device Generating Live Stereo-Soundscapes—Design, Implementation, and Usability Audit

Dominik Osiński [1,*], Marta Łukowska [2], Dag Roar Hjelme [1] and Michał Wierzchoń [2]

[1] Department of Electronic Systems, Norwegian University of Science and Technology, NO-7491 Trondheim, Norway; dag.hjelme@ntnu.no

[2] Consciousness Lab, Institute of Psychology, Jagiellonian University, 30-060 Kraków, Poland; marta.lukowska@doctoral.uj.edu.pl (M.Ł.); michal.wierzchon@uj.edu.pl (M.W.)

* Correspondence: dominik.osinski@ntnu.no

Abstract: The successful development of a system realizing color sonification would enable auditory representation of the visual environment. The primary beneficiary of such a system would be people that cannot directly access visual information—the visually impaired community. Despite the plethora of sensory substitution devices, developing systems that provide intuitive color sonification remains a challenge. This paper presents design considerations, development, and the usability audit of a sensory substitution device that converts spatial color information into soundscapes. The implemented wearable system uses a dedicated color space and continuously generates natural, spatialized sounds based on the information acquired from a camera. We developed two head-mounted prototype devices and two graphical user interface (GUI) versions. The first GUI is dedicated to researchers, and the second has been designed to be easily accessible for visually impaired persons. Finally, we ran fundamental usability tests to evaluate the new spatial color sonification algorithm and to compare the two prototypes. Furthermore, we propose recommendations for the development of the next iteration of the system.

Keywords: Colorophone; sensory substitution; color sonification; multimodal perception; wearable device; assistive device; human–computer interaction

1. Introduction

Visual-to-auditory sensory substitution devices (SSDs) aim to compensate for sensory function loss by delivering information acquired by the visual channel (i.e., via camera or distance sensors) through hearing [1]. Surprisingly, the first electronic SSD of such type was first developed in 1897 by Noiszewski [2]. Nonetheless, there is still no SSD that has been widely accepted broadly by the blind community [3–5]. It contrasts with the recent research results that indicate significant potential for SSDs for non-invasive rehabilitation of the visually impaired [4,6,7] stemming from brain plasticity. According to the multimodal/supramodal brain organization hypothesis [8,9], the human brain operates as a flexible, task-oriented system. Namely, it has been repetitively demonstrated that its organization is more function- than modality-specific. For example, in the case of visual loss, brain areas normally dedicated to visual input processing receive sensory input from other modalities that serve the same function (i.e., spatial cognition) [10]. Importantly, the neuroplastic changes are not restricted to a critical period of brain development and can also occur in adults. This suggests that with proper training, interpretation of the translated sensory information may become intuitive and effortless over time, and a new quality of perceptual experience might be developed [11]. Together with observed progress in the electronic systems field [12], these encouraging results fuel the intensive growth in the development of novel SSDs in recent years [13,14].

However, the color-to-sound coding devices seem relatively unexplored. To the authors' knowledge, only 11 systems realize color sonification [15–25], while only six provide real-time sonification of the acquired color information [18,19,22–25] (see Section 4—Existing Color-to-Sound SSDs for details). One of these was the first version of the Colorophone system [25].

Colorophone

The main goal of the Colorophone project [26] is to develop an affordable, wearable SSD that will enhance perceptual and cognitive capabilities of the visually impaired. We aim to achieve this by providing auditory information about color and distance in an intuitive form. We have performed a preliminary evaluation of the previous version of the Colorophone system conducted on blindfolded, sighted participants. It showed promising results in color and object recognition as well as spatial navigation tasks [25]. Nonetheless, the first prototype had multiple issues regarding its usability and functionality. Firstly, in the previous implementation, we used a standard webcam attached to a headband, which appeared to be bulky. The numerous cables connecting the camera and headphones to the processing unit decreased its usability by reducing users' comfort. Moreover, the generated sounds were artificial sine signals and low-pass filtered white noise that were unpleasant to hear over a prolonged time. Finally, the main functional issue was the fact that the system delivered a nonspatial sensory output by processing the information only from a single area of interest—the focal point of the head-mounted camera. Therefore, we decided to address the above-mentioned disadvantages by developing the next version of the system. Crucially, the new version provides spatial information about colors by sonifying the whole horizontal line of camera pixels. Additionally, we developed a dedicated opponent color space that mimics the human visual system's opponent process, providing more intuitive color categorization, and aims to enhance the auditory color recognition of yellowish colors. Moreover, we improved the appearance and aesthetics of the second version of the system (see Figure 1). A supplementary video example presenting the operation of both versions of the system and a spectrogram of the generated signals can be found in https://youtu.be/fWeKpGMFlmk (accessed on 30 October 2021).

Figure 1. Subsequent versions of the Colorophone system: (**A**) Colorophone 1.0 with a camera mounted on the headband and headphones; (**B**) Colorophone 2.0 with Bose Bluetooth audio sunglasses; (**C**) Colorophone 2.0 with Aftershokz bone-conducting headphones.

2. Design Considerations of SSDs Development

Here, we briefly discuss several design considerations regarding the development of color-to-sound SSDs. We will individually address issues enumerated by Kristjánsson et al. [1]. Additionally, we will comment on several features that can be used to benchmark electronic travel aids (ETAs) presented by Dakopoulos and Bourbakis [27] that can also be helpful while developing new SSDs.

The first design principle states that only critical information about the environment should be conveyed to avoid the risk of sensory overload. Therefore, the designed conversion method should avoid filling up the whole sensory channel; instead, only the chosen parts of the accessible information space should be used.

Another inherent challenge in creating visual-to-auditory SSDs is the mismatch in the information bandwidth between visual and auditory sensory channels. It is estimated that we perceive two or three orders of magnitude more information through vision than through audition [1]. This disproportion is also reflected by the comparison of the number of neural fibers in the optic and auditory nerves. The optic nerve has over one million fibers [28], while the auditory one has over 30,000 fibers [29]. However, the bandwidth of perceptual experience does not reflect the amount of information that can be accessed consciously [30], which is promising information for SSD developers. The human limitation in the amount of consciously attended information can potentially reduce the influence of mismatch in bandwidth. Nonetheless, there still is a firm conviction that SSDs should be task-focused [1]. In our case, the main task for the designed SSD will be color recognition based on auditory color representation [25].

The next design requirement of no interference with other perceptual functions is in our opinion the most difficult to address while designing visual-to-auditory SSDs. It is impossible not to interfere with the acquisition of surrounding sounds while conveying the auditory information via an SSD. However, this interference can be limited by using open-ear headphones and designing the color sonification method to limit the masking effects.

The following requirements are related to design toward usability. The device should be conveniently wearable and easy to operate without using the hands. However, when adjustment in settings is required, the users should be able to straightforwardly change the intensity of the auditory signals according to their demand to perceive environmental or substituted information.

The last design requirement involves a spatiotemporal continuity of coded information. Perception is a continuous process that does not involve a snapshot of the environment [1]. Therefore, the designed method for color sonification should allow for continuous transitions of the auditory information based on spatial and temporal color changes.

3. Why Color?

Although it is challenging to provide a clear definition of what color is [31], it is easier to identify why this attribute of visual perception has an evolutionary function. Color vision provides organisms with important sensory information about the environment that increases their chances of survival [32], by supporting object identification and enhancing the ability for object-ground segmentation [33,34]. Interestingly, the utility of color information for object recognition is greater for medium and low resolutions than for higher resolutions [35]. Still, color is an elusive concept, which cannot be easily described to someone who has never experienced it [31]. A blind person can haptically access information about the shape or distance to the object. However, there is no natural way to access color information via other sensory modalities. This results in the exclusion of visually impaired people from this feature of the perceptual experience, leaving language as the only widely used medium of conveying information about colors.

4. Existing Color-to-Sound SSDs

There are many SSDs that convert visual information to auditory signals [36–43] (see [44] for a detailed review). However, as mentioned before, there are relatively few systems that realize color sonification. Here, we provide short descriptions of color sonification methods implemented in existing systems. Detailed descriptions of the color sonification methods and comparison of experimental results are presented in [25,33]. Information about systems' features, including camera integration, real-time processing and spatialized sound, are shown in Table 1.

ColEnViSon [15] categorizes information about color to one of 10 color categories and associates them with sounds in the following way: red as electric jazz guitar, yellow as a synth drum, brown as a guitar fret noise, orange as a bird tweet, green as a shamisen, blue as a vibraphone, violet as a glockenspiel, black as guitar harmonics, gray as a celesta, and white as a music box. The lighter intensities of the same color are represented as notes on higher scales of the same instruments.

Table 1. Comparison of color-to-sound sensory substitution systems.

System/Author	Camera Integration	Real Time	Spatial Sound
ColEnViSon	No	No	No
HueMusic	No	No	Yes
Musical Vision	No	No	Yes
SoundView	No	Yes	No
Creole	No	Yes	No
EyeMusic	Yes	No	Yes
Sofia Cavaco et al.	Yes	No	Yes
Eyeborg	Yes	Yes	No
KromoPhone	Yes	Yes	No
Colorophone 1.0	Yes	Yes	No
See ColOr	Yes	Yes	Yes

Hue Music [16] categorizes color information as one of the eight distinct hues. Color categorization is implemented using an RGB color model and rounding up to the value of 255 for every value above 127 and rounding down to 0 for every value equal to or above 127. Every hue value is associated with a timbre. Hue values used for timbral associations are red, yellow, green, cyan, blue, magenta, and white. The white color component is represented by silence.

Musical Vision [17] is an image sonification system that uses an RGB color model. Color saturation is coded as volume, and pitch changes represent the spatial location of the pixels. The system reduces color information by discarding the lowest intensity colors and converting RGB values of every pixel into three instruments or chords.

SoundView [18] uses the HSV color model. Grayscale colors are represented by low-pass filtered white noise. The filter's cut-off frequency proportionally depends on brightness levels. Twelve color components are represented by band-pass filtered white noise. The bandwidth of the filters is inversely dependent on color saturation; thus, saturated colors are perceived as tones.

Creole [19] uses CIELUV color space. Color component values are nullified if the amplitude of the color component does not exceed 0.2 of the maximum color value in the processed image. Color intensity is represented by sound loudness. The Creole system represents red as the male vocal vowel sound of "u", yellow as a C major chord (1047, 1319 and 1568 Hz), green as the male vocal vowel sound of "i", blue as a C minor chord (262, 311 and 392 Hz), black as a low-pitched tone of 110 Hz, and white as a high-pitched tone of 3520 Hz. Desaturated colors are represented by band-passed white noise (100 to 3200 Hz).

EyeMusic [20] operates similarly to the vOICe system [36], where the acquired image is processed column by column from left to right, constructing a soundscape. The luminance is coded as loudness, and the vertical position of the processed pixel is associated with pitch changes. However, the sounds used in the EyeMusic system are recordings of musical instruments, and every instrument represents a different color. Red is represented by a reggae organ, yellow by string instruments, green by Rapman's reed, blue as brass instruments, and white as a choir. Various timbres represent the color information, and the pentatonic scale is used for pitch-elevation coding. Only the dominating color for every pixel is played.

Cavaco et al. [21] used the HSV color model. Loudness represents the value, and hue is mapped to pitch (i.e., when the light wave frequency decreases from violet to red, the sound frequency increases). The color saturation is represented by timbre changes from a sinusoid for the lowest saturation to a square wave for the highest saturation.

KromoPhone [23] provides three different color sonification modes, where the default and most advanced is the RGBYW mode. Color intensity is mapped onto the sound volume. Subsequent colors are represented as follows: red as a high-pitch trumpet tone in the right ear, yellow as a high-pitch ukulele tone in the left ear, green as a medium-pitch violin tone in the right ear, and blue as a low-pitch trumpet tone in the left ear. White is represented by high pitch, gray by middle pitch, and black as low pitch—all centrally heard sounds.

Eyeborg [22] codes color saturation as sound volume and continuously transposes light frequencies into sound frequencies. Red is coded by a sound frequency of ≈364 Hz that increases with the light frequency changes up to ≈608 Hz for violet.

The first version of Colorophone [25] uses a dedicated RGBW color space. Color component intensity is coded as loudness. Red is represented by a high-pitch tone of 1600 Hz, green as a middle-pitch tone of 550 Hz, blue as a low-pitch tone of 150 Hz, and white as white noise. The white color component is calculated as a minimum value of RGB color components, which is afterward subtracted from all input RGB values. Black is represented by silence. Amplitudes of color components are perceptually linearized.

The See ColOr system [24] represents colors by using an HSL color model and associates hues in the following way: red is represented by an oboe, orange by a viola, yellow by a pizzicato, green by a flute, cyan by a trumpet, blue by a piano, and purple by a saxophone. The transition between sounds is calculated as a linear relationship between consequent hue values. The pitch of a selected instrument depends on the saturation value. Additionally, darker colors are coded with double bass, while singing voices code brighter colors.

To sum up, there are only four SSDs that provide camera integration and real-time color sonification (two crucial features while considering the use of the system as a blind aid). Only See ColOr [24] uses sound spatialization; however, it does not provide continuous sound output, breaking the spatiotemporal continuity of generated signals. In the next section, we will analyze the possibilities of developing a color sonification method that will enable the generation of intuitive, continuous, and spatialized auditory representation of visual information.

5. Color Sonification

Since the goal of color sonification is to convert information from visual to auditory channels that are inherently different, the necessary preliminary step is to specify the conversion system's purpose. In SSDs used for visual rehabilitation of the blind, the primary function of color sonification algorithms is to provide intuitive information about color by a sound that will enable auditory color perception and recognition. This section builds on the initial design considerations regarding auditory color space development presented in [45].

5.1. Design Considerations Regarding Auditory Color Space

The existing color sonification methods applied in SSDs can be divided into three categories: the first category contains systems that directly associate light frequency with sound frequency [21,22], the second category systems use associations between a predefined color category and the presented sound [20,46] (e.g., the sound of a choir represents the white color). In other words, every color (from a limited palette—usually only a few colors are covered) is represented by an associated sound, which imposes strict color categorization, and sharp transitions between sounds corresponding to different colors. The third category of systems uses basic color components associated with sound components [19,23,25,33]. In such systems, auditory color representation is constructed from many sound components merged into a single auditory stimulus. The devices from the last category provide satisfactory results in experiments related to complex color recognition [25]. Additionally, the approach allows the utilization of the multidimensionality of color experience in the auditory color representation. Therefore, we decided to use the last approach while designing the new color sonification method.

5.1.1. Psychophysics

Since sight and hearing show different psychophysical characteristics, we implemented an inverted Stevens's power law [47] for the auditory channel. It compensates for the nonlinear volume perception of the human auditory system. The information about the color intensity is preprocessed by the inverted Stevens's power law function, which then is annulated by the influence of the human auditory system.

5.1.2. Cross-Modal Correspondences

Cross-modal correspondences are natural associations between different sensory modalities, such as bright objects being loud [48]. The usage of cross-modal correspondences enhances the performance of auditory color recognition [19]. Although finding a universal mapping of various sensory modalities remains ambiguous, we can utilize existing research results as a guideline in designing the color sonification method. The first intuitive mapping between a color component and a sound component would be mapping the intensity of the color stimuli to the intensity of the sound stimuli. More intensive colors will be associated with higher volume sounds. We chose to associate color components with corresponding sound frequencies on the basis of the pitch–chroma relationship described in [19].

5.1.3. Number of Color Components

When selecting the number of color components to be represented by a sonification algorithm, we should bear in mind that if this number is too large, it will be difficult for a naïve user to remember and recognize all the color–sounds associations. However, if the number is too low, a user will not have the necessary variety in the auditory signal to be able to recognize a color change. Our preliminary tests indicated that the RGBW color space [25] allowed satisfactory auditory color recognition of 14 tested colors (black, white, red, pale red, green, pale green, blue, pale blue, yellow, pale yellow, violet, pale violet, cyan, and pale cyan), but the recognition of colors near yellow (orange, olive green) appeared challenging. Although these two colors are visually perceived as saliently different, their sound representations were perceived as very similar. In addition to improving discrimination between yellowish colors, adding the yellow color component into the developed color space has other advantages. Firstly, red, yellow, green, and blue are assumed to be elementary colors called unique hues, and the subjective appearance of any other color can be composed of these unique hues [49]. Moreover, the yellow component is central in opponent process theory [50]; thus, the yellow–blue axis is present in many advanced color spaces. Therefore, we consider the yellow color component to be necessary for our color sonification design. Black remains a unique color component because the information about this color, which effectively means lack of any light, can be analogously conveyed by silence—the lack of any sound. The proposed color space of five color components plus black strongly reminds of color component definition from the Natural Color System (NCS) [51] that is entirely based on the phenomenology of human perception.

5.2. Color Spaces

NCS is one of the various color spaces that define the conventions of coding information about color with numerical values. CIELAB and CIELUV are often used, where uniform color spaces are based on the opponent process theory [31]. However, neither of the mentioned color spaces have focal colors as color axes. CIELAB does not have focal red, blue, or green anywhere close to the corresponding color axes, and CIELUV has the most significant deviation from the axes for green, yellow, and red color components [50]. By focal colors, Kuehni [52] defines the ideal representatives of a given basic color name.

While designing auditory color space based on previous considerations, we need to use a color space based on opponent process theory, where color axes are as close as possible to focal red, yellow, green, and blue. We propose to call the color space equipped with the features described above as RYGBW, where letters represent the following color components: red, yellow, green, blue, and white.

5.2.1. RYGBW Auditory Color Space

The RYGBW color space is constructed based on the RGB color model by calculating the W component as a minimum value of all RGB components that is subtracted from the original RGB values. Thereafter, the Y component is dependent on the amplitudes of R and G components in a way that makes the transition of color components between red and yellow, and yellow and green similar to the transitions for other unique hues. To visualize color component variability, we plot respective RYGBW values for color transitions presented using an HSL color model. The transitions between fully saturated colors for the whole hue spectrum are presented in Figure 2. According to the design idea, the yellow component becomes a new dimension in the constructed color space, similar to red, green, and blue.

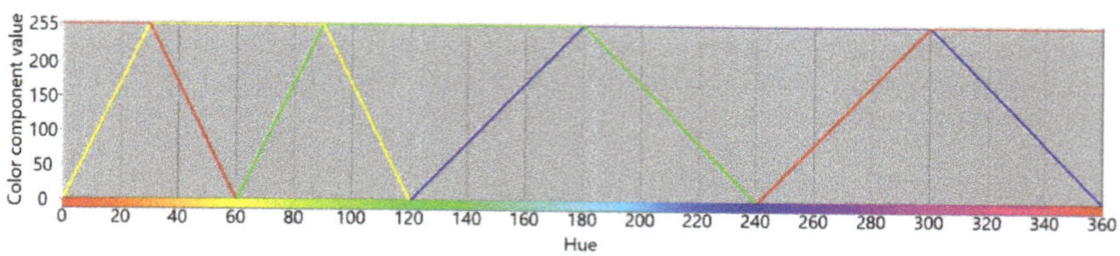

Figure 2. Transitions between chromatic color components in RYGBW color space.

Figures 3–6 present transition profiles for individual color components from black, through fully saturated color, to white. When the white color component increases, the value of other color components decrease respectively. Figure 7 presents color transition for non-saturated colors from black through gray to white.

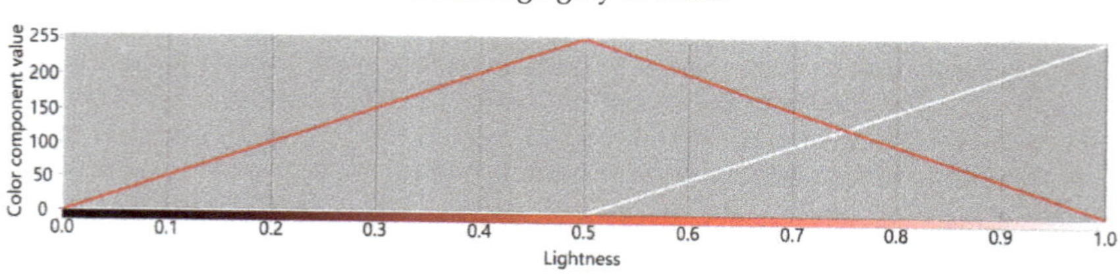

Figure 3. Example of the color transition for a single-color component from black through red to white in RYGBW color space.

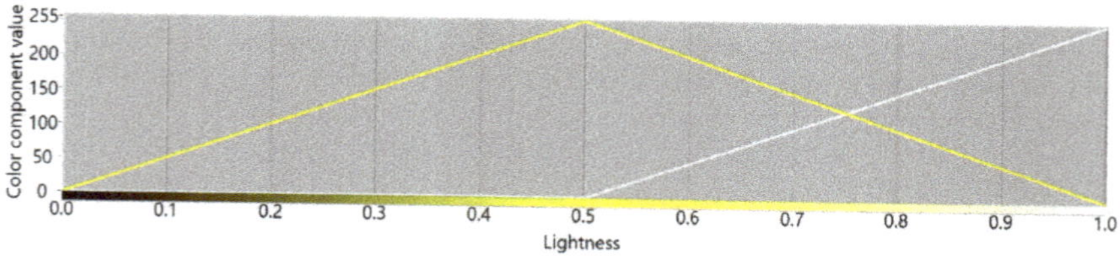

Figure 4. Example of the color transition for a single-color component from black through yellow to white in RYGBW color space.

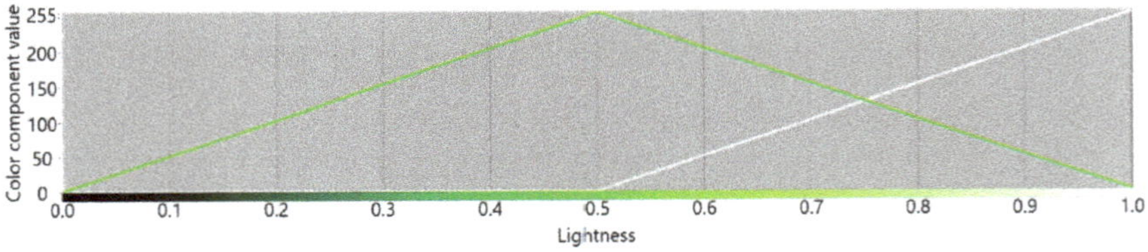

Figure 5. Example of the color transition for a single-color component from black through green to white in RYGBW color space.

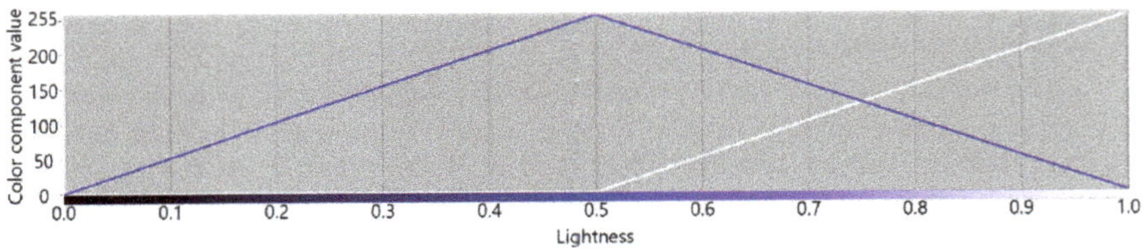

Figure 6. Example of the color transition for a single-color component from black through blue to white in RYGBW color space.

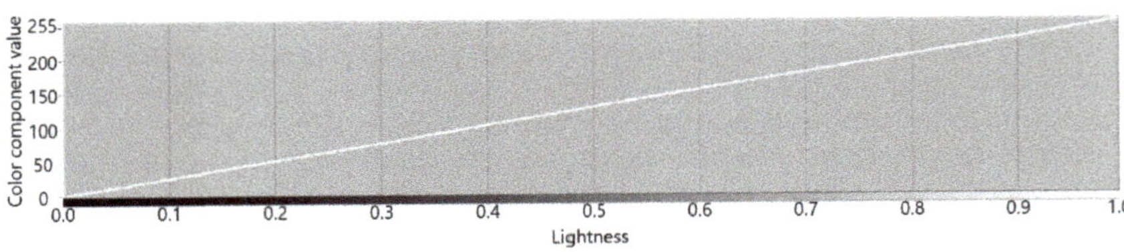

Figure 7. Example of the color transition for a single-color component from black to white in RYGBW color space.

5.2.2. Sounds Associated with Color Components

While choosing sounds corresponding to colors, we used the following guidelines:

- The sounds should:
  - Be pleasant for the user [25];
  - Be calibrated in amplitude corresponding to the maximal color intensity to provide equal loudness for every sound component;
  - Have higher difference in frequency than the critical bands to avoid masking effects [53];
  - >Be preferably perceptually equally spaced in pitch [54]; and,
  - Be associated with colors on the basis of chosen cross-modal correspondences (i.e., blue—low pitch, green—middle-low pitch, yellow—middle-high pitch, red—high pitch).
- White should be coded by a sound with no characteristic primary frequency such as white noise or rainfall.

Since we know which sound pairs will be presented together, we can choose to simultaneously present only dissonant pairs of sounds, which positively influences the recognition of sound components [55]. Importantly, to preserve the sound continuity, we considered only musical instruments that allow for seamless looping of the used sound samples such as the violin or trumpet. The ceiling frequency was 1027 Hz to maintain a high resolution in sound localization possibilities [56,57] (for details, see Section 5.3.1 Sound Localization) and avoid high-pitch sounds that are perceived as unpleasant [58]. The lowest used frequency was empirically chosen to avoid excessive vibration of bone-conducting headphones. Based on these considerations, we decided to choose the associations between the color and sound components presented in Table 2.

Table 2. Chosen associations between color and sound components.

Color Component	Sound Frequency (Hz)	Note	Sound Type
Red	1027	C6	Musical instruments
Yellow	647	E5	
Green	408	G#4	
Blue	256	C4	
White	-	-	Rainfall

5.3. Spatial Color Sonification Algorithm

The human ability of sound source localization opens a possibility for the development of a spatialized color sonification algorithm. This may be realized by applying the sonification method described above to code the color information from a larger number of areas of interest (zones) located on the processed image into auditory signals. However, it requires the parallel computation of color to sound conversions for multiple zones using high-quality digital waveforms and applying functions necessary for sound spatialization.

5.3.1. Sound Localization

When a sound reaches our ears, we use subtle differences in sound timing, intensity, and spectral composition to determine sound source location [59]. The angles describing the sound source location in the polar coordinates system are called azimuth for the horizontal plane and elevation for the vertical plane. The difference in distance between our ears and the sound source results in interaural time difference (ITD), and the shadowing effect produced by the head causes interaural intensity difference (IID). The ITD and IID together are called binaural localization cues [59]. The ITD is the dominant binaural cue—it is a major cue for the localization of low-frequency sounds, but it also contributes to high-frequency sound localization [60]. It is assumed that the diffraction effect of an average human head is negligible for sound waves of frequencies below 1 kHz and that IID is too small to facilitate sound localization for frequencies below 1500 Hz [59]. The minimum audible angle (MAA) parameter is used to investigate the human sound localization ability. The MAA defines the smallest perceptible difference in the position of a sound. In the sighted population, it has been demonstrated that for wideband stimuli and low-frequency tones presented in the frontal position in the horizontal plane, the MAA is on the order of 1° to 2° [59,61]. However, the MAA depends on the position of the sound source in the horizontal plane and a sound frequency. Importantly, the ability to localize a sound source decreases rapidly for frequencies between 1050 and 2500 Hz [56].

Moreover, sound localization is more precise in the frontal (i.e., when one is located frontally to the sound sources) as compared to the lateral (i.e., when one is located laterally to the sound sources) position. Namely, it has been demonstrated that an average error in absolute localization for a broadband sound source is about 5° for the frontal and about 20° for the lateral position [59].

For moving sound sources, the minimum audible movement angle (MAMA) is used to determine the limits of sound localization abilities. It has been demonstrated that changes in sound frequency similarly influence both MAA and MAMA [62]. Namely, the MAMA is smaller for signals below 1050 Hz than for higher frequencies [57]. However, the relationship is nonlinear (i.e., in the range between 250 and 1050 Hz, it takes a U-shape and increases for frequencies above 1050 Hz; for details, see [56]). Another experiment that investigated MAMA for a broadband noise source moving at the velocity of 20°/s showed MAMA values on the order of 2° for azimuth angles in the range of 0–40° and 4° for the angle of 80° [63].

The highest base sound frequency used in the Colorophone system is 1027 Hz. It has been chosen to meet the limit of accurate sound localization based on ITD [56] and dynamic spatial resolution [57]. Here, we propose the spatialized sound implementation based only on ITD. For calculating ITD, we use a frequency-independent model of a wave propagating around a sphere expressed by Woodworth's formula [64]:

$$ITD = \frac{a}{c}(sin\theta + \theta), \tag{1}$$

where a is the radius of the sphere, c is the speed of sound (343 m/s), and θ is the lateral angle. For the average head radius value, we used 87 mm from an estimation of a spherical head model based on anthropometry [64]. It is essential in the context of SSD development that sound localization ability is preserved while using bone-conductive headphones [65].

5.3.2. Zone Size Determination

Based on the camera's viewing angle and image resolution, sound generation parameters, and limits of spatial auditory resolution, we can calculate the minimal zone size and consequently the number of sonification zones. Since the number of zones influences the calculational load of the system, it is important to configure the zones' sizes and positions in a way that sounds coming from adjacent zones should be potentially distinguishable. For low-frequency tones arriving from the frontal position, the MAA corresponds to an ITD differences of 10–20 µs. The chosen sound sampling frequency will determine the minimum temporal resolution to time difference corresponding to one sample. The standard sampling frequency recommended by the Audio Engineering Society for professional digital audio is 44,100 Hz [66]. The time difference between samples for 44,100 Hz frequency is ≈23 µs, which matches the ITD difference corresponding to MAA. For the horizontal camera resolution of 640 pixels and 90° field of view (FoV), the zone size corresponding to a 1° viewing angle is 7.1 pixels. However, taking into account MAMA, which will probably be more suitable considering enactive head movements, the minimum zone size can correspond to 2–3° of the camera's viewing angle (i.e., ≈14–21 px for the camera's FoV of 90°). Nonetheless, it is essential to remember that MAMA increases for larger azimuth values; therefore, while using a wider FoV, the minimal zone size will increase to 4–6°, which corresponds to ≈28–43 pixels. To sum up, we chose the minimal zone size of 14 px for the central azimuth values and gradually increased the zones' sizes for the higher azimuth values corresponding to lateral zones.

6. The Colorophone 2.0 SSD

Here, we propose two implementations of the Colorophone system that consist of a Bluetooth camera and headphones, and a processing unit in the form of a Windows tablet (see Figure 8). We have also implemented two software interfaces—one designed for researchers and the second one for visually impaired users. Both versions use the same sonification algorithm that performs visual data acquisition, data processing, and sound generation.

Figure 8. Two realizations of the Colorophone 2.0 system together with the processing unit in the form of a Windows tablet.

6.1. Wearable Prototypes

We have built two wearable versions of the system: one based on Bose Frames audio sunglasses [67] and the second based on Aftershokz Aeropex [68] bone-conducting headphones. The comparison of the relevant headphone parameters is presented in Table 3.

Table 3. Comparison of the two open-ear headphones used for prototyping.

Headphones	Bose	Aftershokz
Weight (g)	50	26
Battery life (h)	Up to 5.5	8
Charging time 0–100% (min)	60	90
Waterproof (IP)	No (IPX2)	Yes (IP67)
Retail price ($)	249.95	159.95

Both cameras are equipped with an OmniVision OV2735 image sensor and a USB-C 2.0 interface. The field of view of both cameras is 90°. The camera used in the prototype based on the Bose Frames uses a two-point magnetic connection (on the left side), while the camera used in Aftershokz Aeropex is mounted on the right side of the headphones with a flexible arm and tape (see Figure 9). The applied mounting solutions enable free interaction with functional buttons of both prototypes.

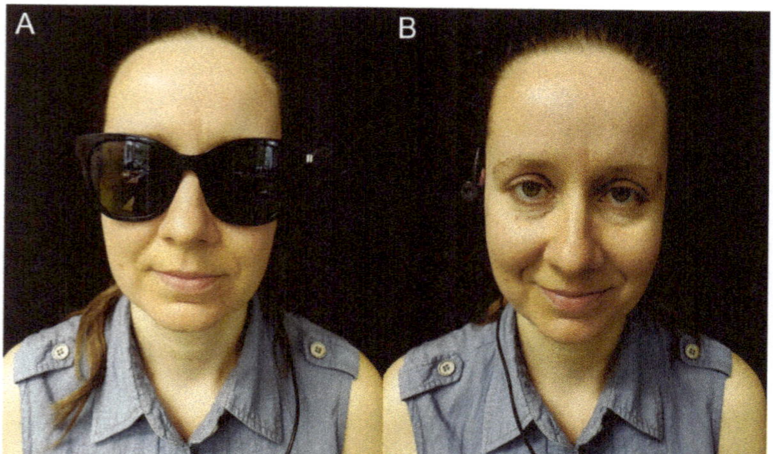

Figure 9. Two realizations of a wearable camera-headphone prototype: (**A**) Bose Frames; (**B**) Aftershokz Aeropex.

6.2. Processing Unit

The processing unit is an HP 608 Pro tablet equipped with an Intel Atom x5-Z8500 with Intel HD Graphics (1.44 GHz, up to 2.24 GHz using Intel Burst Technology, 2 MB cache, and 4 cores), 4 GB LPDDR3-1600 SDRAM, 64 GB embedded Multi Media Card (eMMC), 7.86-inch diagonal capacitive multi-touch, FHD QXGA BrightView WLED UWVA (2048 × 1536), and a 21 Wh lithium–polymer battery. The tablet's external dimensions are 137 mm by 207 mm by 8.35 mm, and it weighs 420 g. The installed operating system is Windows 10.

6.3. Software

The software has been developed using LabVIEW 2020 by National Instruments [69] with an add-on Vision Development Module. LabVIEW is a programming environment that allows for relatively easy system development and integration of various peripheral devices. The developed system acquires images via a USB camera, codes visual data into waveforms, and outputs sound via Bluetooth headphones. The functional block diagram showing the operations of the system is presented in Figure 10.

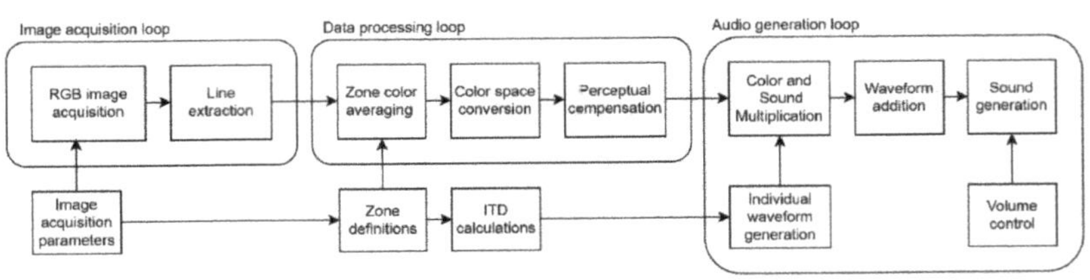

Figure 10. Functional block diagram of the sonification algorithm implemented in the Colorophone 2.0 system.

The system operates in three parallel loops responsible for image acquisition, data processing, and audio generation. Although dependent on the data transfer between them, these loops communicate in a way that ensures stable operation of the whole system. For example, if the image acquisition loop automatically reduces the frame rate in response to poor lighting conditions, it does not interrupt the data processing loop's function and, consequently, the audio generation loop. During the start of the system, all loops are initialized with configuration parameters that influence various loop functions.

6.3.1. Image Acquisition

The first loop in our processing pipeline is responsible for the continuous acquisition of RGB images from an external USB camera. Before starting the loop operations, a dedicated function identifies possible video operation modes of a connected camera. Then, a chosen video mode of 640 × 480 pixels is used to initialize the connection with the camera. Then, images are acquired continuously at the rate of 30 frames per second. Then, the image is converted to an array of pixel values, and a horizontal line is extracted from the array and sent via a local variable to the next data processing loop.

6.3.2. Data Processing

The second data processing loop reads the configuration data calculated based on image acquisition parameters and zone definitions. These data are used for setting zone boundaries, determining which pixels should be included for every zone. Then, the averaging of color information for every zone is performed. The output RGB information is converted into RYGBW color space, and a compensation ensuring perceptual linearity in the auditory channel is applied. The current implementation data processing loop calculates color values in parallel for 15 zones and sends the information about RYGBW parameters for every zone to the following audio generation loop.

6.3.3. Audio Generation

The last loop generates auditory signals based on ITD calculations and processes the information about color data received from the previous loop. During program initialization, .wav files containing sound samples corresponding to every color component are loaded into the memory. Then, 80 individual waveforms are generated based on data received from the ITD calculation function. These waveforms are looped in order to preserve sound continuity. Then, the amplitudes of waveform values are multiplied by their respective color component amplitudes for every zone and by the volume control value. The single audio generation loop iteration takes 30 ms; thus, the maximum information processing time of the whole system is 60 ms. Every sound loop iteration creates a new soundscape that is sent to the default Windows audio output device.

6.3.4. Interface for Researchers

Figure 11 presents the dedicated graphical user interface (GUI) for researchers. The acquired image is presented together with a visualization of the averaged color information for every processed zone. The GUI allows access to multiple configurative functions, such as choosing the camera and switching between color sonification modes for nonspatial and spatial processing.

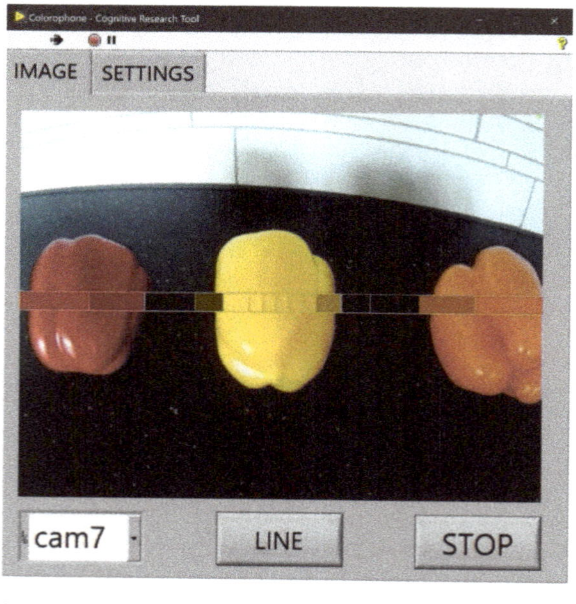

Figure 11. The Colorophone interface for researchers—the main image tab of the GUI.

The settings tab (Figure 12) contains more configuration options. Here, one can flip left and right channels. This option can be used for non-standard camera mounting that involves rotation of the camera. It is also possible to choose various video parameters of camera operation by identifying the desired image resolution from the list of available video modes and using a video mode string to extract the chosen option. Another variable enables the user to set the camera's FoV and define sizes of individual zones. The control sum of the pixels from all zones is displayed to prevent errors in zone sizes definition. The program also allows for the choice of sound samples used individually for every channel by the sonification algorithm.

6.3.5. Interface for Blind Users

The second developed interface is implemented both in the form of a graphical as well as an auditory user interface. The GUI has been designed to reflect the need for a high-contrast display and contains a limited number of buttons (Figure 13). *START* and *STOP* buttons are used for turning on and off the sonification process—the *LINE* button switches between the point and zone processing mode. The mode change is also reflected in the appearance of the color box above the *LINE* button (i.e., it switches between presenting the color information for nonspatial and spatial modes). The auditory user interface operates by using interaction cues—a user explores the whole screen haptically, and when they touch a button localization on the screen, a voice command reads the button name. The second touch of the button activates the chosen option, and a voice command of "*Going to ...* " is played for the user together with the button's label. For example, when a user slides their finger over the *LINE* button, the first played message is "*Line*"; after a click, the *LINE* mode is activated, and the message "*Going to line*" is played. The system also generates speech-based error messages. When the camera gets disconnected during the operation of the system, the user receives the following message: "*Houston, we have a problem. Check the camera cable and restart the app*".

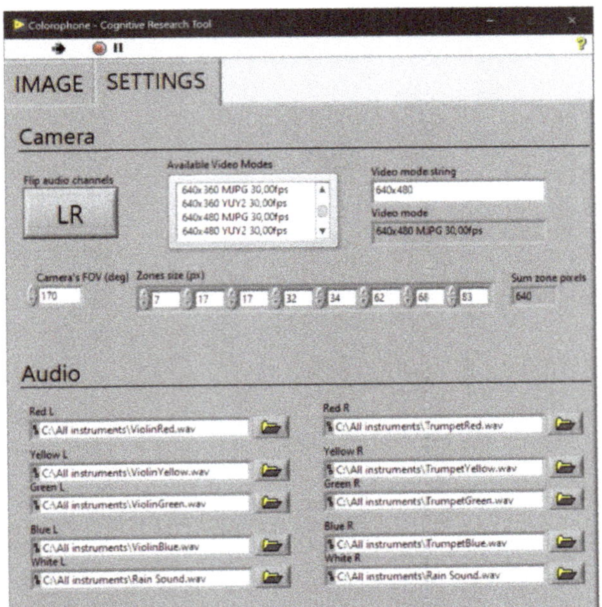

Figure 12. The Colorophone interface for researchers—settings tab of the GUI.

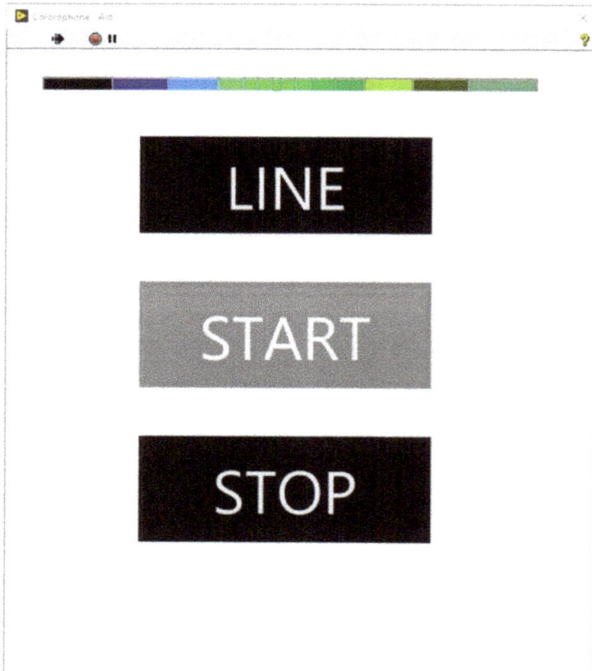

Figure 13. The Colorophone high-contrast GUI.

6.4. Applied System Settings

For the purpose of the system evaluation, it has been set up in the following way: violin sounds were associated with the colors on the left side of the image, while trumpet sounds coded colors on the right side of the image. Using two different instruments provides an extra cue regarding sound spatialization. The color of the central zone was associated with simultaneously played violin and trumpet sounds. The sound of rainfall represented white. To investigate the possibility of using the whole angular range of sound localization, the camera FoV was set to 170°, so the camera's image was "stretched" in the auditory domain to cover almost the whole azimuth sound variability. We defined 15 zone sizes ranging from 14 to 83 pixels (i.e., one central and seven on the left and right side).

7. Evaluation of the System

A usability audit was conducted to initially evaluate the functionality of the Colorophone 2.0 and prepare for in-depth, multidimensional usability tests with visually impaired users. The audit was conducted by an interdisciplinary research team consisting of a UX specialist, a cognitive science researcher, and a qualitative studies researcher. All three specialists were sighted and gave written consent to participate in the audit. Importantly, the UX specialist has never had contact with the device before, whereas the two other experts were experienced in sensory substitution research. We aimed to compare the two prototypes, assess the intuitiveness of the color-to-sound mapping, analyze differences in the two sonification area modes (i.e., nonspatial and spatial), track sensorimotor contingencies (i.e., the regularities in how sensory stimulation depends on the activity of the perceiver/user [70]) required by the device, and evaluate the system usability to solve everyday life tasks (i.e., natural object recognition, color identification, reaching, and locomotion). The audit was conducted in three steps, starting with a free exploration phase when the specialists were allowed to interact with the system without a defined goal. Then, a task testing the Colorophone usability in the peripersonal (i.e., within hand reach) space was administered. In the task, a user was positioned in front of a table, where a few pairs of colorful objects of the same size were located (see Figure 14A). The user was asked to find an object of a given color, point on its location, and grasp it. All three specialists underwent the task under four conditions resulting from the combination of the two variables: prototype (Bose vs. Aftershokz) and sonification area mode (nonspatial vs. spatial). Then, the second task, testing the system usability to solve tasks in extrapersonal (i.e., behind hand reach) space was administered (see Figure 14B). The task was to find a green wall in a corridor and then recognize the shape of an object located on the wall. Given that the specialists did not undergo any formal training of the Colorophone's usage before the audit, we can conclude that they managed to fulfill all tasks to a satisfactory level. Namely, after a while, they were able to recognize and find the object of a given color in peripersonal space and navigate slowly in the indoor environment. Given users' expertise level and sample size (i.e., the three specialists), we employed a qualitative approach to identify potential problems and propose an adequate improvement. Here, we present the main conclusions from the audit with recommendations for further studies.

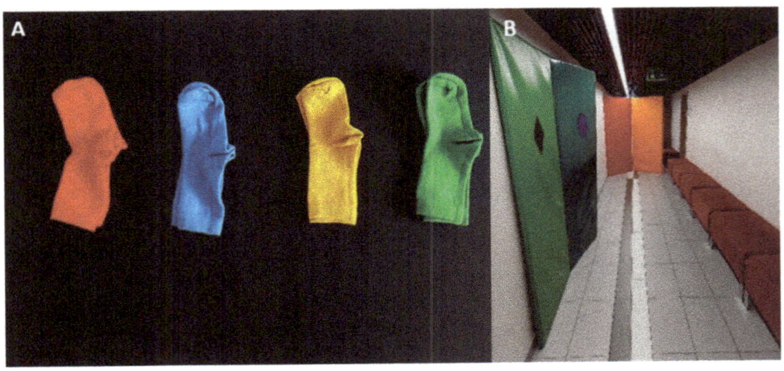

Figure 14. Setup used in the two tasks of the usability audit: (**A**) peripersonal space task (color recognition, object localization, pointing, and grasping); (**B**) extrapersonal space task (colorful object localization, route planning, locomotion, object's shape recognition).

Firstly, after the exploration and the first task, we concluded that the Bose-based prototype is more suitable because the Aftershokz implementation is unstable (i.e., it was moving during exploratory head movements, constantly changing the position of the sampling camera attached to the headphones). In effect, users cannot develop efficient sensorimotor contingencies because of the variability of the sampling device in relation to their body and external space. Namely, they were unable to correctly interpret where the signal was coming from and, as a result, could not locate objects in space based on the sonified color information. However, we found that the Bose headphones are restricting access to external sounds more than the Aftershokz bone-conducting ones. Therefore, we suggest to either develop the Bose-based prototype or to use the Aftershokz but with the camera attached separately to avoid instability.

Secondly, we found that the applied transformation of the camera field of view into a sound space was confusing. Namely, the camera's FoV is 90°, which is converted to 170° in the sound space. This transformation creates serious confusion as to where an object is located in physical space. Additionally, in both prototypes, the camera is located on the head side, which causes even more confusion concerning objects localization in reference to the sampling device and body. Both of the above-mentioned problems have implications for the ability to locate sound sources in space solely on the basis of information provided by the Colorophone. Moreover, it poses a risk of a long-lasting egocentric reference frame (ERF) recalibration as in case of a hemispatial neglect; patients suffering from this medical condition omit objects located in their left visual field due to an altered central body axis sensation. In our case, the ERF recalibration may result in a sensation as if the central body axis is shifted (i.e., translated or rotated) toward the camera location [71]. More specifically, a user learns the relationship between the object's location in space and the location of the sound in the auditory space provided by the Colorophone (i.e., acquires new sensorimotor contingencies). Later, when we remove the device, the egocentric reference frame is recalibrated, and the user may have a problem with a correct localization of stimuli based on auditory information that is no longer mediated by the device. This is because they continue to apply the correction acquired over the training with the device. This is potentially dangerous, as it can have a long-lasting effect as a prismatic adaptation (i.e., egocentric reference frame recalibration treatment applied in the hemispatial neglect patients) [72,73]. Importantly, in case of visually impaired users, a readaptation to an adequate egocentric reference frame might be difficult due to restricted access to visual information [74]. Apart from confusion and the risk of egocentric reference frame recalibration, the current spatial transformation and the lateral camera location might cause headaches and nausea, as it was reported already after two hours of using the device by all three users. However, the reported aversiveness of the white color sound representation might also contribute to the observed headache. Nonetheless, we suggest to change the visual-to-auditory spatial transformation algorithm to provide proper correspondence between an objects' location and its auditory space (i.e., the auditory space should be restricted exactly to the camera FoV). Additionally, we recommend locating the sampling device (i.e., camera) on a central body axis (e.g., in the middle of the forehead or between eyes). Optionally, a hand-held version of the camera could be enabled to avoid neck muscle tension after long-term head-mounted Colorophone prototype usage.

Thirdly, the comparison of the nonspatial (i.e., one sonification zone) and spatial (i.e., multiple sonification zones) modes revealed that they might serve different functions. Namely, the former could be more efficiently used in peripersonal (i.e., close) space tasks (e.g., object location within reach and grasping), whereas the latter could be more efficiently used in extrapersonal (i.e., far) space tasks (e.g., object location behind reach, route planning, and navigation). Moreover, the modes seem to complement each other to support some functions; for example, in the indoor navigation, the spatial mode could be used to have a glimpse of the whole surroundings and locate an object that one wants to move toward, and then, the nonspatial mode might serve as a 'rope' enabling to keep direction during locomotion toward the chosen direction. However, the current implementation of the spatial mode should be reconsidered to account better for the auditory spatial resolution of visually impaired users. This is because in the audit, the users found it difficult to identify which sound is heard in which spatial zone. It might be potentially even harder for visually impaired users, since some of them could demonstrate lower auditory spatial resolution (i.e., minimum audible angle) [75–78]. Therefore, further studies concerning the optimal number and size of the zones in the spatial mode should be conducted. Additionally, individual differences in MAA should be considered to customize the multiple zones mode, and further investigations should identify which functions are best supported by which mode, as well as how the modes can complement each other to support everyday life activities of visually impaired users.

Finally, we identified several issues concerning color-to-sound mapping. The users found the sound representation of white color distracting and potentially aversive (i.e., motivating to avoid exploration of bright environments). Given the omnipresence of white color in natural and indoor environments, we should consider choosing a less attention-consuming and more pleasant sound corresponding to white color or reduce the device sensitivity to this component. Moreover, black representation as silence was also found as problematic, especially when trying to recognize shapes of black objects. So, adding a sonified black color representation should be considered or introducing a function that would reverse white and black color auditory representations. Additionally, we recognized a problem with variability introduced by changing lighting (i.e., introduced by user's shade, light source changes, or by observation angle changes). In the visual modality, top–down processes provide color constancy [79], but the current implementation of the Colorophone sonification algorithm does not account sufficiently for the lighting-induced variability. Importantly, it significantly impacts the ability to identify objects based on the sonified color information, since the way a given object sounds is radically affected by the lighting context. The lack of a color stability problem has to be solved to support the Colorophone-mediated color cognition.

In summary, the current implementation of the Colorophone system seems promising as a tool to support both color and spatial cognition. However, the above-mentioned issues concerning visual-to-auditory spatial transformation, the sampling device location, white and black color sound representation, and color stability have to be solved to increase Colorophone usability. Importantly, we should bear in mind that the audit was conducted without systematic training, and the experts cannot be perceived as SSD super-users. It is possible that a prolonged practice with the Colorophone will solve some of the above-mentioned problems by decreasing cognitive load, embodying the device, and enabling an intuitive usage strategy [80]. Therefore, studies with well-trained visually impaired users in their natural environment might shed more light on Colorophone usability.

8. Conclusions and Further Work

We have successfully developed and initially evaluated a wearable, real-time, spatialized color sonification system. The system uses a dedicated opponent color space that mimics some of the low-level functions of the human visual system. The device generates continuous, rich, and relatively pleasant (i.e., as compared to other SSDs) soundscapes. In contrast to the previous version of the system, the current version provides spatial color sonification for 15 zones based on a pixel line covering the whole horizontal FoV of the camera. Sine signals and white noise has been replaced with more pleasant sounds of instruments and rainfall. We have proposed two new realizations of the device that use mini cameras, providing a relatively discreet and modern look. The system allowed users to fulfill the tasks of finding an object of a given color in peripersonal space and navigating slowly in the indoor environment without conducting prior systematic training.

The disadvantages of the system's current implementation include issues connected to the lateral mounting of the camera and scaling of the camera's FoV, the camera's inability to accurately compensate for various light conditions that results in object's colors inconstancies and potential identification problem. Importantly, the system's implementation does not provide information about the distance to the objects. However, this issue has been addressed by developing an additional Colorophone visual echolocation function introduced in [81].

Further work will include addressing the problems pointed out by usability audit and further tests of the next version of the system with visually impaired users. We will focus primarily on improving the spatial sonification algorithm. Firstly, we will consider results concerning the auditory spatial resolution of visually impaired users [75–78] to define the number of zones and their sizes in the spatial mode. Secondly, we will implement assumptions of the auditory scene analysis [82,83] to define how many auditory streams can be experienced as separate at the same time with the spatialized version of the Colorophone. Finally, we will evaluate several configurations of the multiple sonification zone mode with the visually impaired users to determine the multiple sonification zones mode. Additionally, we aim to develop a more informative black sound representation and less aversive white sound representation to increase the signal-to-noise ratio and improve colorful object detection on a bright or dark background, as well as to account for an illumination difference between daytime and nighttime. Moreover, we plan to investigate the whole Interactive System (composed of User, Environment, and Technology, according to the Personalized Inclusive SSDs Design model that is currently under development). So, not only will we focus on the device development, but also, we will consider how to adapt the user's environment and account for individual differences between users to enhance Colorophone usability for everyday life tasks.

Author Contributions: Conceptualization, D.O. and D.R.H.; methodology, D.O., D.R.H. and M.Ł.; software, D.O.; validation, D.O. and M.Ł.; formal analysis, D.O. and M.Ł.; investigation, D.O. and M.Ł.; resources, D.O. and M.W.; data curation, D.O. and M.Ł.; writing—original draft preparation, D.O. and M.Ł.; writing—review and editing, M.Ł, D.R.H. and M.W.; visualization, D.O.; project administration, D.O. and M.Ł.; funding acquisition, D.O. and M.W. All authors have read and agreed to the published version of the manuscript.

Funding: This research was supported by the National Science Centre, Poland, grant OPUS (2016/23/B/HS6/00275) given to Michał Wierzchoń and internal NTNU's Kompetanseløft grant.

Institutional Review Board Statement: The study was conducted according to the guidelines of the Declaration of Helsinki, and approved by the Ethics Committee of Institute of Psychology.

Informed Consent Statement: Informed consent was obtained from all subjects involved in the study.

Data Availability Statement: Data sharing not applicable.

Acknowledgments: We would like to thank Justyna Janik and Weronika Kałwak for performing the usability audit, Falk Morawitz for developing sound samples and Peter Svensson for advices regarding the implementation of the sound localization algorithm.

Conflicts of Interest: The authors declare no conflict of interest.

References

1. Kristjánsson, Á.; Moldoveanu, A.; Jóhannesson, Ó.I.; Balan, O.; Spagnol, S.; Valgeirsdóttir, V.V.; Unnthorsson, R. Designing Sensory-Substitution Devices: Principles, Pitfalls and Potential. *RNN* **2016**, *34*, 769–787. [CrossRef] [PubMed]
2. Starkiewicz, W.; Kuliszewski, T. The 80-Channel Elektroftalm. In Proceedings of the International Congress on Technology and Blindness, New York, NY, USA, 1 January 1963; Volume 1, p. 157.
3. Capp, M.; Picton, P. The Optophone: An Electronic Blind Aid. *Eng. Sci. Educ. J.* **2000**, *9*, 137–143. [CrossRef]
4. Maidenbaum, S.; Abboud, S.; Amedi, A. Sensory Substitution: Closing the Gap between Basic Research and Widespread Practical Visual Rehabilitation. *Neurosci. Biobehav. Rev.* **2014**, *41*, 3–15. [CrossRef]
5. Elli, G.V.; Benetti, S.; Collignon, O. Is There a Future for Sensory Substitution Outside Academic Laboratories? *Multisens. Res.* **2014**, *27*, 271–291. [CrossRef]
6. Proulx, M.J.; Ptito, M.; Amedi, A. Multisensory Integration, Sensory Substitution and Visual Rehabilitation. *Neurosci. Biobehav. Rev.* **2014**, *41*, 1–2. [CrossRef]
7. Reich, L.; Maidenbaum, S.; Amedi, A. The Brain as a Flexible Task Machine: Implications for Visual Rehabilitation Using Noninvasive vs. Invasive Approaches. *Curr. Opin. Neurol.* **2012**, *25*, 86–95. [CrossRef]
8. Ricciardi, E.; Bonino, D.; Pellegrini, S.; Pietrini, P. Mind the Blind Brain to Understand the Sighted One! Is There a Supramodal Cortical Functional Architecture? *Neurosci. Biobehav. Rev.* **2014**, *41*, 64–77. [CrossRef] [PubMed]
9. Rosenblum, L.D.; Dias, J.W.; Dorsi, J. The Supramodal Brain: Implications for Auditory Perception. *J. Cogn. Psychol.* **2017**, *29*, 65–87. [CrossRef]
10. Chebat, D.-R.; Schneider, F.C.; Ptito, M. Spatial Competence and Brain Plasticity in Congenital Blindness via Sensory Substitution Devices. *Front. Neurosci.* **2020**, *14*, 1–18. [CrossRef]
11. Jóhannesson, Ó.; Balan, O.; Unnthorsson, R.; Moldoveanu, A.; Kristjánsson, Á. The Sound of Vision Project: On the Feasibility of an Audio-Haptic Representation of the Environment, for the Visually Impaired. *Brain Sci.* **2016**, *6*, 20. [CrossRef]
12. Csapó, Á.; Wersényi, G.; Nagy, H.; Stockman, T. A Survey of Assistive Technologies and Applications for Blind Users on Mobile Platforms: A Review and Foundation for Research. *J. Multimodal. User Interfaces* **2015**, *9*, 275–286. [CrossRef]
13. Bhowmick, A.; Hazarika, S.M. An Insight into Assistive Technology for the Visually Impaired and Blind People: State-of-the-Art and Future Trends. *J. Multimodal. User Interfaces* **2017**, *11*, 149–172. [CrossRef]
14. Tapu, R.; Mocanu, B.; Zaharia, T. Wearable Assistive Devices for Visually Impaired: A State of the Art Survey. *Pattern Recognit. Lett.* **2020**, *137*, 37–52. [CrossRef]
15. Ancuti, C.; Ancuti, C.; Bekaert, P. *ColEnViSon: Color Enhanced Visual Sonifier-A Polyphonic Audio Texture and Salient Scene Analysis*; SciTePress-Science and and Technology Publications: Lisboa, Portugal, 2009; pp. 566–572.
16. Payling, D.; Mills, S.; Howle, T. Hue Music-Creating Timbral Soundscapes from Coloured Pictures. In Proceedings of the 13th International Conference on Auditory Display, Montréal, QC, Canada, 26–29 June 2007; pp. 91–97.
17. Polo, A.; Sevillano, X. Musical Vision: An Interactive Bio-Inspired Sonification Tool to Convert Images into Music. *J. Multimodal User Interfaces* **2019**, *13*, 231–243. [CrossRef]

18. Van den Doel, K. Soundview: Sensing Color Images by Kinesthetic Audio. In Proceedings of the 2003 International Conference on Auditory Display, Boston, MA, USA, 6–9 July 2003; pp. 303–306.
19. Hamilton-Fletcher, G.; Wright, T.D.; Ward, J. Cross-Modal Correspondences Enhance Performance on a Colour-to-Sound Sensory Substitution Device. *Multisens. Res.* **2016**, *29*, 337–363. [CrossRef] [PubMed]
20. Abboud, S.; Hanassy, S.; Levy-Tzedek, S.; Maidenbaum, S.; Amedi, A. EyeMusic: Introducing a "Visual" Colorful Experience for the Blind Using Auditory Sensory Substitution. *Restor. Neurol. Neurosci.* **2014**, *32*, 247–257. [CrossRef]
21. Cavaco, S.; Henriques, J.T.; Mengucci, M.; Correia, N.; Medeiros, F. Color Sonification for the Visually Impaired. *Procedia Technol.* **2013**, *9*, 1048–1057. [CrossRef]
22. Eyeborg. 2021. Available online: https://en.wikipedia.org/wiki/Eyeborg (accessed on 9 March 2021).
23. Capalbo, Z.; Glenney, B. Hearing Color: Radical Pluralistic Realism and SSDs. In Proceedings of the AP-CAP, Tokyo, Japan, 1 October 2009; pp. 135–140.
24. Bologna, G.; Deville, B.; Pun, T. On the Use of the Auditory Pathway to Represent Image Scenes in Real-Time. *Neurocomputing* **2009**, *72*, 839–849. [CrossRef]
25. Osinski, D.; Hjelme, D.R. A Sensory Substitution Device Inspired by the Human Visual System. In Proceedings of the 2018 11th International Conference on Human System Interaction (HSI), Gdansk, Poland, 4–6 July 2018; pp. 186–192.
26. Colorophone. Available online: https://www.colorophone.com (accessed on 29 March 2021).
27. Dakopoulos, D.; Bourbakis, N.G. Wearable Obstacle Avoidance Electronic Travel Aids for Blind: A Survey. *IEEE Trans. Syst. Man Cybern. C* **2010**, *40*, 25–35. [CrossRef]
28. Jonas, J.B.; Müller-Bergh, J.A.; Schlötzer-Schrehardt, U.M.; Naumann, G.O. Histomorphometry of the Human Optic Nerve. *Investig. Ophthalmol. Vis. Sci.* **1990**, *31*, 736–744.
29. Spoendlin, H.; Schrott, A. Analysis of the Human Auditory Nerve. *Hear. Res.* **1989**, *43*, 25–38. [CrossRef]
30. Cohen, M.A.; Dennett, D.C.; Kanwisher, N. What Is the Bandwidth of Perceptual Experience? *Trends Cogn. Sci.* **2016**, *20*, 324–335. [CrossRef]
31. Fairchild, M.D. Color Appearance Models. In *Wiley-IS&T Series in Imaging Science and Technology*, 2nd ed.; Wiley: Chichester, West Sussex, UK; Hoboken, NJ, USA, 2005; ISBN 978-0-470-01216-1.
32. Gerl, E.J.; Morris, M.R. The Causes and Consequences of Color Vision. *Evo. Edu. Outreach* **2008**, *1*, 476–486. [CrossRef]
33. Hamilton-Fletcher, G.; Ward, J. Representing Colour Through Hearing and Touch in Sensory Substitution Devices. *Multisens. Res.* **2013**, *26*, 503–532. [CrossRef]
34. Kremers, J.; Baraas, R.C.; Marshall, N.J. (Eds.) *Human Color Vision*; Springer International Publishing: Cham, Switzerland, 2016; ISBN 978-3-319-44976-0.
35. Torralba, A. How Many Pixels Make an Image? *Vis. Neurosci.* **2009**, *26*, 123–131. [CrossRef] [PubMed]
36. Meijer, P.B.L. An Experimental System for Auditory Image Representations. *IEEE Trans. Biomed. Eng.* **1992**, *39*, 112–121. [CrossRef] [PubMed]
37. Strumillo, P. Electronic Interfaces Aiding the Visually Impaired in Environmental Access, Mobility and Navigation. In Proceedings of the 3rd International Conference on Human System Interaction, Rzeszow, Poland, 13–15 May 2010; pp. 17–24.
38. Bujacz, M. Representing 3D Scenes through Spatial Audio in an Electronic Travel Aid for the Blind. Ph.D. Thesis, Lodz University of Technology (Politechnika Łódzka), Łódź, Poland, 2010; p. 978.
39. Rodríguez, A.; Yebes, J.J.; Alcantarilla, P.F.; Bergasa, L.M.; Almazán, J.; Cela, A. Assisting the Visually Impaired: Obstacle Detection and Warning System by Acoustic Feedback. *Sensors* **2012**, *12*, 17476–17496. [CrossRef]
40. Ribeiro, F.; Florêncio, D.; Chou, P.A.; Zhang, Z. Auditory Augmented Reality: Object Sonification for the Visually Impaired. In Proceedings of the 2012 IEEE 14th International Workshop on Multimedia Signal Processing (MMSP), Banff, AB, Canada, 17–19 September 2012; pp. 319–324.
41. Balakrishnan, G.; Sainarayanan, G.; Nagarajan, R.; Yaacob, S. A Stereo Image Processing System for Visually Impaired. *Int. J. Comput. Inf. Eng.* **2008**, *2*, 2794–2803.
42. Saez, J.M.; Escolano, F.; Lozano, M.A. Aerial Obstacle Detection with 3-D Mobile Devices. *IEEE J. Biomed. Health Inform.* **2015**, *19*, 74–80. [CrossRef]
43. Caraiman, S.; Zvoristeanu, O.; Burlacu, A.; Herghelegiu, P. Stereo Vision Based Sensory Substitution for the Visually Impaired. *Sensors* **2019**, *19*, 2771. [CrossRef]
44. Bujacz, M.; Strumiłło, P. Sonification: Review of Auditory Display Solutions in Electronic Travel Aids for the Blind. *Arch. Acoust.* **2016**, *41*, 401–414. [CrossRef]
45. Osinski, D.; Bizon, P.; Midtfjord, H.; Wierzchon, M.; Hjelme, D.R. Designing Auditory Color Space for Color Sonification Systems. In Proceedings of the Proceedings of the 25th International Conference on Auditory Display (ICAD 2019), Newcastle upon Tyne, UK, 23–27 June 2019; pp. 331–334.
46. Bologna, G.; Deville, B.; Pun, T. Sonification of Color and Depth in a Mobility Aid for Blind People. In Proceedings of the Proceedings of the 16th International Conference on Auditory Display (ICAD2010), Washington, DC, USA, 9–15 June 2010; p. 6.
47. Stevens, S.S. On the Psychological Law. *Psychol. Rev.* **1957**, *64*, 153–181. [CrossRef]
48. Spence, C. Crossmodal Correspondences: A Tutorial Review. *Atten. Percept. Psychophys* **2011**, *73*, 971–995. [CrossRef] [PubMed]
49. Witzel, C.; Gegenfurtner, K.R. Are Red, Yellow, Green, and Blue Perceptual Categories? *Vis. Res.* **2018**, *151*, 152–163. [CrossRef] [PubMed]

50. Moroney, N. *The Opposite of Green Is Purple?* SPIE: San Jose, CA, USA, 2009; p. 72410N.
51. NCS Colour-Universal Language for Colour Commmunication. Available online: https://ncscolour.com/ (accessed on 6 March 2021).
52. Kuehni, R.G. Focal Colors and Unique Hues. *Color Res. Appl.* **2001**, *26*, 171–172. [CrossRef]
53. Fastl, H.; Zwicker, E. *Psychoacoustics: Facts and Models*; Springer: New York, NY, USA, 2007; ISBN 978-3-540-68888-4.
54. Stevens, S.S.; Volkmann, J.; Newman, E.B. A Scale for the Measurement of the Psychological Magnitude Pitch. *J. Acoust. Soc. Am.* **1937**, *8*, 185–190. [CrossRef]
55. Brown, D.J.; Simpson, A.J.R.; Proulx, M.J. Auditory Scene Analysis and Sonified Visual Images. Does Consonance Negatively Impact on Object Formation When Using Complex Sonified Stimuli? *Front. Psychol.* **2015**, *6*, 1–9. [CrossRef]
56. Mills, A.W. On the Minimum Audible Angle. *J. Acoust. Soc. Am.* **1958**, *30*, 237–246. [CrossRef]
57. Perrott, D.R.; Tucker, J. Minimum Audible Movement Angle as a Function of Signal Frequency and the Velocity of the Source. *J. Acoust. Soc. Am.* **1988**, *83*, 1522–1527. [CrossRef]
58. Kumar, S.; Forster, H.M.; Bailey, P.; Griffiths, T.D. Mapping Unpleasantness of Sounds to Their Auditory Representation. *J. Acoust. Soc. Am.* **2008**, *124*, 3810–3817. [CrossRef]
59. Letowski, T.R.; Letowski, S.T. *Auditory Spatial Perception: Auditory Localization*; Defense Technical Information Center: Fort Belvoir, VA, USA, 2012.
60. Macpherson, E.A.; Middlebrooks, J.C. Listener Weighting of Cues for Lateral Angle: The Duplex Theory of Sound Localization Revisited. *J. Acoust. Soc. Am.* **2002**, *111*, 2219. [CrossRef]
61. Perrott, D.R.; Saberi, K. Minimum Audible Angle Thresholds for Sources Varying in Both Elevation and Azimuth. *J. Acoust. Soc. Am.* **1990**, *87*, 1728–1731. [CrossRef]
62. Chandler, D.W.; Grantham, D.W. Minimum Audible Movement Angle in the Horizontal Plane as a Function of Stimulus Frequency and Bandwidth, Source Azimuth, and Velocity. *J. Acoust. Soc. Am.* **1992**, *91*, 1624–1636. [CrossRef]
63. Strybel, T.Z.; Manllgas, C.L.; Perrott, D.R. Minimum Audible Movement Angle as a Function of the Azimuth and Elevation of the Source. *Hum. Factors* **1992**, *34*, 267–275. [CrossRef]
64. Algazi, R.; Avendano, C.; Duda, R.O. Estimation of a Spherical-Head Model from Anthropometry. *J. Aud. Eng. Soc.* **2001**, *49*, 472–479.
65. MacDonald, J.A.; Henry, P.P.; Letowski, T.R. Spatial Audio through a Bone Conduction Interface. *Int. J. Audiol.* **2006**, *45*, 595–599. [CrossRef] [PubMed]
66. Cabot, R.C. *Preferred Sampling Frequencies for Applications Employing Pulse-Code Modulation*; Audio Engineering Society: New York, NY, USA, 2018.
67. Bluetooth Audio Sunglasses | Bose. Available online: https://www.bose.com/en_us/products/frames.html (accessed on 9 March 2021).
68. Aeropex. Available online: https://us.aftershokz.com/products/aeropex (accessed on 16 March 2021).
69. National Instruments. Available online: https://www.ni.com/ (accessed on 18 March 2021).
70. Degenaar, J.; O'Regan, J.K. Sensorimotor Theory of Consciousness. *Scholarpedia* **2015**, *10*, 4952. [CrossRef]
71. Karnath, H.-O.; Dieterich, M. Spatial Neglect—A Vestibular Disorder? *Brain* **2006**, *129*, 293–305. [CrossRef] [PubMed]
72. Jacquin-Courtois, S.; O'Shea, J.; Luauté, J.; Pisella, L.; Revol, P.; Mizuno, K.; Rode, G.; Rossetti, Y. Rehabilitation of Spatial Neglect by Prism Adaptation. *Neurosci. Biobehav. Rev.* **2013**, *37*, 594–609. [CrossRef]
73. Welch, R.B. Adaptation to Prism-Displaced Vision: The Importance of Target-Pointing. *Percept. Psychophys.* **1969**, *5*, 305–309. [CrossRef]
74. Iachini, T. Does Blindness Affect Egocentric and Allocentric Frames of Reference in Small and Large Scale Spaces? *Behav. Brain Res.* **2014**, *9*, 73–81. [CrossRef]
75. Aggius-Vella, E.; Kolarik, A.J.; Gori, M.; Cirstea, S.; Campus, C.; Moore, B.C.; Pardhan, S. Comparison of Auditory Spatial Bisection and Minimum Audible Angle in Front, Lateral, and Back Space. *Sci. Rep.* **2020**, *10*, 1–9. [CrossRef]
76. Gori, M.; Sandini, G.; Martinoli, C.; Burr, D.C. Impairment of Auditory Spatial Localization in Congenitally Blind Human Subjects. *Brain* **2014**, *137*, 288–293. [CrossRef] [PubMed]
77. Vercillo, T.; Burr, D.; Gori, M. Early Visual Deprivation Severely Compromises the Auditory Sense of Space in Congenitally Blind Children. *Dev. Psychol.* **2016**, *52*, 847–853. [CrossRef]
78. Voss, P.; Lassonde, M.; Gougoux, F.; Fortin, M.; Guillemot, J.-P.; Lepore, F. Early- and Late-Onset Blind Individuals Show Supra-Normal Auditory Abilities in Far-Space. *Curr. Biol.* **2004**, *14*, 1734–1738. [CrossRef] [PubMed]
79. Foster, D.H. Color Constancy. *Vis. Res.* **2011**, *51*, 674–700. [CrossRef] [PubMed]
80. Alfaro, A.; Bernabeu, Á.; Agulló, C.; Parra, J.; Fernández, E. Hearing Colors: An Example of Brain Plasticity. *Front. Syst. Neurosci.* **2015**, *9*. [CrossRef]
81. Bizoń-Angov, P.; Osiński, D.; Wierzchoń, M.; Konieczny, J. Visual Echolocation Concept for the Colorophone Sensory Substitution Device Using Virtual Reality. *Sensors* **2021**, *21*, 237. [CrossRef]
82. Bregman, A.S. *Auditory Scene Analysis: The Perceptual Organization of Sound*; MIT Press: Cambridge, MA, USA, 1994; ISBN 978-0-262-52195-6.
83. Moore, B.C.J.; Gockel, H.E. Properties of Auditory Stream Formation. *Philos Trans. R. Soc. Lond. B Biol. Sci.* **2012**, *367*, 919–931. [CrossRef]

sensors

MDPI

Article

Blindness and the Reliability of Downwards Sensors to Avoid Obstacles: A Study with the EyeCane

Maxime Bleau [1], Samuel Paré [1], Ismaël Djerourou [1], Daniel R. Chebat [2,3], Ron Kupers [1,4] and Maurice Ptito [1,4,*]

1 École D'optométrie, Université de Montréal, Montréal, QC H3T 1P1, Canada; maxime.bleau.1@umontreal.ca (M.B.); samuel.pare.1@umontreal.ca (S.P.); ismael.djerourou@umontreal.ca (I.D.); endogonidia@gmail.com (R.K.)
2 Visual and Cognitive Neuroscience Laboratory (VCN Lab.), Department of Psychology, Faculty of Social Sciences and Humanities, Ariel University, Ari'el 40700, Israel; danielc@ariel.ac.il
3 Navigation and Accessibility Research Center of Ariel University (NARCA), Ari'el 40700, Israel
4 Department of Neuroscience, University of Copenhagen, 2200 Copenhagen, Denmark
* Correspondence: maurice.ptito@umontreal.ca; Tel.: +45-28552324

Abstract: Vision loss has dramatic repercussions on the quality of life of affected people, particularly with respect to their orientation and mobility. Many devices are available to help blind people to navigate in their environment. The EyeCane is a recently developed electronic travel aid (ETA) that is inexpensive and easy to use, allowing for the detection of obstacles lying ahead within a 2 m range. The goal of this study was to investigate the potential of the EyeCane as a primary aid for spatial navigation. Three groups of participants were recruited: early blind, late blind, and sighted. They were first trained with the EyeCane and then tested in a life-size obstacle course with four obstacles types: cube, door, post, and step. Subjects were requested to cross the corridor while detecting, identifying, and avoiding the obstacles. Each participant had to perform 12 runs with 12 different obstacles configurations. All participants were able to learn quickly to use the EyeCane and successfully complete all trials. Amongst the various obstacles, the step appeared to prove the hardest to detect and resulted in more collisions. Although the EyeCane was effective for detecting obstacles lying ahead, its downward sensor did not reliably detect those on the ground, rendering downward obstacles more hazardous for navigation.

Keywords: navigation; blindness; sensory substitution; EyeCane; obstacle detection; avoidance; collision

1. Introduction

While navigating in an environment, humans rely on visual information to identify obstacles, evaluate distances, and create a mental map of their surroundings. Thus, the loss of visual input as occurs in blindness has a proportionally greater effect on navigational abilities and independence as compared to other senses [1]. To overcome this deficit in safe locomotion, blind individuals must learn to use many aids and tools to obtain environmental information that is necessary for diverse everyday tasks such as wayfinding and circumventing obstacles [2]. As the most widespread of these mobility aids, the white cane functions as an extension of the hand and arm to enable obstacle detection and to furnish information about ground textures and level changes (i.e., drop-offs, steps, and curbs) encountered during locomotion. However, the white cane does not generally provide information about obstacles positioned higher than the user's pelvis, thus leaving the upper body at risk of collisions and injuries [3,4] (Figure 1A). This significant limitation considerably impedes safety, which can be discouraging to blind individuals and ultimately leads to social isolation [5]. There is therefore a pressing need to develop and explore new technologies that could potentially improve their safety and autonomy in daily travels. One promising area of research is sensory substitution, which aims to convey visual information to blind individuals through touch and sound with sensory substitution devices (SSD) [6]. Many different SSDs have been designed to substitute for vision, while others provide guidance that is more task-based, meaning that they offer a specific aspect of visual information (e.g., color, shape, distance or letters) strictly pertinent to preforming a certain task (e.g., wayfinding, obstacle circumvention, or reading).

For instance, the tongue display unit (TDU) is a tactile-to-vision SSD that, in a laboratory environment, allows blind individuals to discriminate shapes [7,8], movement [9], pathways [10], and letters [11,12], while enabling users to detect and avoid obstacles in a life-size obstacle course [13]. However, the visually impaired community has not adopted the TDU, since the abundant information given by the device is complex and requires the user to attend constantly to maintain an adequate level of performance, which quickly leads to exhaustion (see cognitive load problem in [6,14,15]). Therefore, devices like minimalist SSDs and electronic travel aids (ETA) that convey a simpler signal could maintain safety while limiting fatigue [2,6].

ETA devices are equipped with sensors (ultrasonic, infrared, or electromagnetic), radars, or cameras to capture information about the environment and convey encoded signals to blind users by tactile (vibration) or auditory stimulation [2,16–18]. While ETAs fall within the general category of SSDs, they mostly provide simple feedback that is strictly relevant for navigation, such as the presence of obstacles in the path of travel, and their distance and location. Since the 20th century, researchers have developed multiple ETAs and other SSDs as mobility aids to assist blind individuals [2]. These devices generally qualify as either primary tools (that can be used independently from the white cane) [19] or secondary tools (that must be used in conjunction with the white cane) [20,21]. However, the broader blind population has not adopted such aids due to many factors such as training requirements, high cost, and low portability [6,22].

To circumvent these shortcomings, the EyeCane was developed as a torch-like ETA equipped with narrow field of view (FOV) infrared sensors that capture distance information about obstacles and convey it to the user's hand through vibrations. This gives the possibility to "feel" the immediate environment without touching it directly. Several studies that evaluated the EyeCane's potential in rehabilitation have concluded that its users were able easily to estimate distances, navigate, and detect and avoid obstacles in real [23] and virtual [24,25] environments after only a brief training.

In a recent study [26], the EyeCane was adapted with two narrow FOV infrared sensors with a 1.5 m sensing range (Figure 1B) to test its reliability as a primary and secondary aid to tackle the safety challenge posed by waist-up obstacles. The authors of that study stated the hypothesis that the EyeCane might suffice as a reliable primary mobility aid if it were equipped with a third sensor pointing toward the ground (Figure 1D), thus replacing the need for a physical cane for detecting foot-obstacles. The goal of the present study was to investigate the reliability of such a downward sensor. For this purpose, we used a single narrow FOV infrared sensor with a 2 m sensing range pointed toward the ground (Figure 1C), and we assessed its reliability in a life-size obstacle course presenting risks of collision, tripping, and falling. Furthermore, we compared the EyeCane navigational capacities of late (LB)- and early-blind individuals (EB) to blindfolded sighted controls (SC). We hypothesized that smaller obstacles at ground level would prove more difficult to detect for the three groups of subjects, and that EB would be more adept than the two other groups at learning to navigate the obstacle course.

2. Materials and Methods

2.1. Participants and Ethics

Participants were recruited in Montreal (Canada) through the database of the Harland Sanders Research Chair in Vision Neuroscience, and in Denmark through the database of the BRAINlab at the University of Copenhagen. The experiment took place at the School of Optometry at the Université de Montréal and the University of Copenhagen.

Ten EB (mean age: 44 ± 12 years; 2 females and 8 males) and nine LB (mean age: 48 ± 14 years; 4 females and 5 males) were recruited in this study (Table 1).

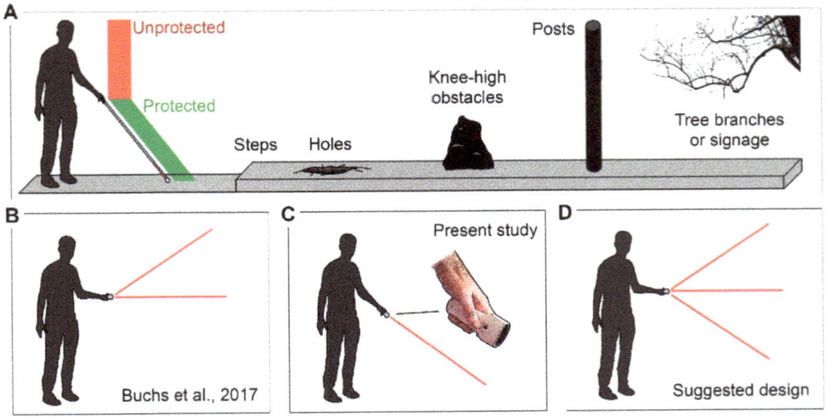

Figure 1. Schematics of (**A**) an individual with the white cane facing daily-life obstacles, the cane protects him from low obstacles; however, the individual is at risk of head injury with hanging obstacles like tree branches and signage; (**B**) an individual handling the EyeCane design used in [26]; (**C**) our experimental design with the single-sensor EyeCane in a downward directed manner and (**D**) an individual handling the three-sensors EyeCane design suggested in [26].

Table 1. Blind participants' characteristics.

Participant	Age	Sex	Age of Onset of Blindness	Blindness Duration Index	Cause of Blindness	Residual Perception
LB1	55	F [1]	24	0.56	Retinitis pigmentosa	yes
LB2	25	M [2]	17	0.32	Retinitis pigmentosa	-
LB3	70	M	38	0.46	Meningitis	-
LB4	38	F	20	0.64	Retinal cancer	-
LB5	46	M	40	0.13	Meningitis	-
LB6	56	F	20	0.47	Retinal cancer	-
LB7	47	F	22	0.53	Diabetic retinopathy	-
LB8	44	F	17	0.61	Glaucoma	-
LB9	59	F	57	0.03	Retinitis pigmentosa	yes
EB1	48	M	Perinatal	-	Retinopathy of prematurity	-
EB2	33	M	Perinatal	-	Retinopathy of prematurity	-
EB3	63	M	Perinatal	-	Retinopathy of prematurity	-
EB4	54	M	Perinatal	-	Retinopathy of prematurity	-
EB5	56	M	Perinatal	-	Retinopathy of prematurity	-
EB6	36	M	Perinatal	-	Retinopathy of prematurity	-
EB7	31	M	Perinatal	-	Retinopathy of prematurity	-
EB8	40	M	Perinatal	-	Retinopathy of prematurity	-
EB9	33	F	Perinatal	-	Retinopathy of prematurity	-
EB10	51	M	Perinatal	-	Meningitis	-

[1] Female, [2] Male.

We also included 10 sighted controls (SC) with normal vision (mean age: 44 ± 13 years; 4 females and 4 males), who were age- and sex-matched to the blind participants. SC were blindfolded throughout the experiment. All blind participants were expert users of the white cane, and two had a guide dog. To evaluate the influence of experience-dependent plasticity in the LB group, we calculated the blindness duration index (BDI) according to the formula "(age–age onset blindness)/age" (as described in [27]). The BDI score can range from 0 to 1, expressing the proportion of his life that a person has been blind, with low scores indicating recent onset of blindness and high scores longer duration of blindness. The average BDI was 0.42 ± 0.20 (range: 0.03 to 0.64) while the mean onset of blindness was 28.2 ± 13.7 years. Participants had no associated neuropathy that could affect their navigational performance or mental representation. Before starting the experiment, participants completed a questionnaire regarding their blindness and spatialization abilities and signed a consent form. The protocol was approved by the Clinical Research Ethics Committee of the University of Montreal (CERC-19-097-P) and by the local ethics committee of the University of Copenhagen (Region Hovestaden; Protocol nr: H-6-2013-004) and was conducted in accordance with the Declaration of Helsinki.

2.2. Apparatus

The EyeCane is a small (4 × 5 × 13 cm) and light-weight (∼100 g) hand-held mobility device with a form similar to a flashlight (Figures 1 and 2) and a long battery life (up to 24 h use, simple to charge) [23]. Equipped with an infrared emitter and sensor (Sharp GP2Y0A02YK0F), it emits a narrow light beam (<5°) in the direction at which it is aimed and detects the reflected signal. The EyeCane then determines the distance to the hit object and translates the information into vibration, encoded with varying intensity levels. Therefore, when an obstruction is detected at a range of 20 to 150 cm, the device vibrates in the user's palm, and its intensity is inversely proportional to the distance of the obstacle: the closer the obstacle, the higher the intensity of the vibration. The full specification details have been described previously [23], and the sensors were previously shown to work on different materials and in various lighting conditions [23,28,29].

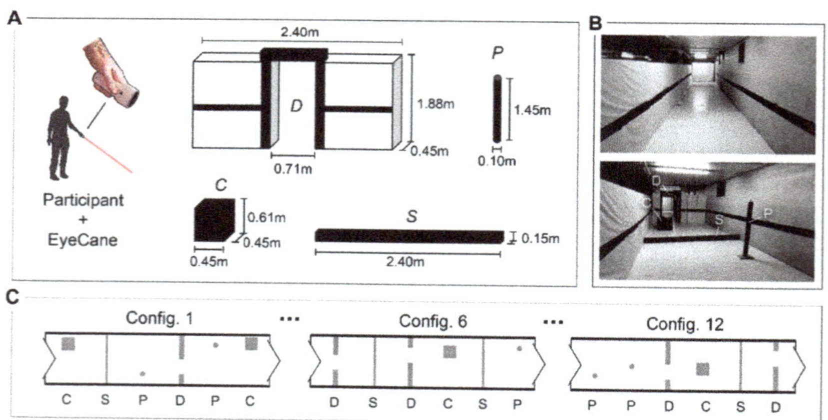

Figure 2. The experimental setup and procedure. (**A**) A participant handling the EyeCane in a downward directed manner; four type of obstacles: D= door, P = post, C = cube, and S = step. (**B**) At the top, a photograph of the empty obstacle course and at the bottom an example of obstacle placement. (**C**) Three schematic representations of the twelve different trial configurations.

2.3. Experimental Procedure

The experiment consisted of three parts: the training phase, the test phase, and the post-test phase, when the participants' experience and level of satisfaction were assessed.

2.3.1. Obstacle Course

The experiments were conducted in a life-size obstacle course simulating daily-life situations encountered during outdoor and indoor travel. This obstacle course consisted of a hallway measuring 21 m long and 2.4 m wide, equipped with differently sized obstacles to evaluate detection, avoidance, and identification performances with the use of the EyeCane as a standalone aid (Figure 2). To minimize the risk of injury, all obstacles were constructed from cardboard and foam.

Four types of obstacles were designed for the experiment. These included "cube" (height: 0.61 m; width: 0.45 m; depth: 0.45 m), "door frame" (height: 1.88 m; depth: 0.45 m; door width: 0.71 m; total width: 2.4 m), "step" (height: 0.15 m; depth: 0.15 m; width: 2.4 m), and "post" (height: 1.45 m; diameter: 0.10 m) (Figure 2). These four types of obstacles were chosen to represent a range of scenarios with differences in floor surfaces and other obstacles in the path of travel. The "cube" represented large knee-high obstacles; the "door frame" represented narrow passages such as door frames and any space between two obstacles or between two other pedestrians; the "step" represented possible changes in the floor denivelation (i.e., steps, curbs, and sidewalks); and the "post" represented thin obstacles often found on the sidewalk such as signage.

2.3.2. Training Phase

All participants underwent the same training procedures. They were first verbally introduced about the EyeCane's principle of operation and then familiarized with the device in an otherwise empty room containing a single obstacle, i.e., a cardboard tower (0.4 × 0.4 × 2 m). This familiarization phase served to introduce the concepts of obstacle recognition, size, and distance estimation, while being guided and unguided by the experimenters. Participants were taught that object recognition is possible by scanning the object with the device (side-to-side and up–down) and detecting its edges to gain information about its shape. The participants then underwent a simulated detection and avoidance task in the experimental walkway (21 × 2.4 m), with placement of three cardboard towers at 3 m apart on the longitudinal axis and randomly positioned on the horizontal axis.

The participants were then introduced to the scanning technique used for the experiment. They were taught how to scan the environment with the device pointed toward the ground in front of them such that they felt a constant, low-intensity vibration emitted from the device. This technique ensures that users always detect the ground and are thus alerted to any changes in the floor surface or the space in front of them. In fact, the arm movements required for this technique were closely related to the "two-point touch technique" that blind individuals learn in the context of white cane orientation and mobility (O&M) lessons [30,31]. The goal of this specific technique was to simulate the use of a third sensor (pointed toward the ground) while isolating it from the two other sensors' signals. Therefore, our experimental design allows us to assess the reliability of this specific sensor without any added complexity and, thus, to assess the suggested three-sensors EyeCane's suitability as a primary mobility aid.

Finally, the participants were familiarized with the four types of obstacles used in the experiment. During this phase, the participants used the scanning technique and were guided by the experimenter toward each obstacle. They were taught to detect and identify each of the four types of obstacles. The complete training phase (device familiarization, scanning technique, and obstacles familiarization) averaged 15 min for most participants.

2.3.3. Test Phase

All participants navigated the same 12 configurations of the obstacle test in random order. For each configuration, six obstacles of random type were placed as in Figure 2, thus comprising a total of 18 encounters for each obstacle.

The assigned task was to cross the corridor as quickly as possible while detecting, identifying, and avoiding obstacles. The participants were monitored by two experimenters for safety reasons and data collection. Object *detection, identification* and *avoidance* are distinct processes that follow each other sequentially and serve the same goal: ensuring the individual's safety during navigation. Detection is the first step toward attaining the navigation goal as it allows the individual to gain awareness about the presence of an obstacle in the path of travel, to adjust his/her pace for safety, and to anticipate contact or avoidance, thus decreasing the risk of a dangerous collision. The *identification* process occurs when the individual gains information about the nature and dimensions of an object.

Avoidance is the culmination of both processes, occurring when the individual has to plan a deviation in his/her path of travel according to the obstacle's position and dimensions, to successfully execute this new path, and finally to regain the initial track. Therefore, to serve effectively as a primary mobility aid, the EyeCane must prove reliable for obstacle detection, identification, and avoidance, serving for each type of obstacles, especially those at ground level (i.e., the task's "step" obstacle, potholes, and curbs). Another important factor for efficiency and reliability is the amount of time needed to detect, identify, avoid obstacles and complete the course; thus, we also measured crossing time.

2.4. Statistical Analysis

The collected data consisted of the average crossing time and indices of three types of performance: obstacle detection, avoidance, and identification. Since avoidance performance also counted those obstacles that participants did not encounter (being too small or too peripheral to their track), we therefore separately evaluated participants' performances in avoiding detected obstacles. Collision data was also calculated as the opposite scalar of avoidance performance.

Data were analyzed using JASP, an open-source graphical program for statistical analysis developed by the University of Amsterdam [32]. Two-way ANCOVA tests corrected for age and sex, or the nonparametric equivalent Kruskal–Wallis test, were used to determine the effect of group (EB, LB, and SC) on time and performance data. We then verified the effects of obstacle type ("cube", "door frame", "step", and "post") and group (EB, LB, and SC) on detection, as well as their interaction on collision rates, using two-way ANCOVA corrected for age and sex. *Post hoc* T-tests (or Mann–Whitney) with Bonferroni correction were performed to identify any significant differences. Results are presented as mean ± SD.

3. Results

Crossing time: In terms of the average time to cross the obstacle course through the 12 trials, EB were faster with 154.5 ± 39.6 s than LB with 240.7 ± 117.2 s and SC with 304.9 ± 143.8 s (Figure 3A). The ANCOVA corrected for age and sex test indicated that there was a statistically significant difference in crossing times between groups ($F(2,22) = 7.290$, $p = 0.004$, $\eta^2 = 0.384$) but failed to show an effect of age or sex ($p > 0.05$). Post hoc comparison with Bonferroni correction revealed that EB were significantly faster than SC ($p = 0.003$), but no differences were found between EB and LB or between LB and SC ($p > 0.05$).

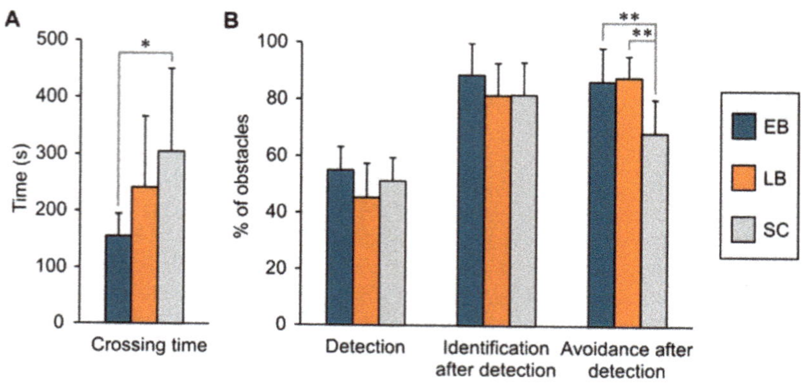

Figure 3. The average crossing time (**A**), and the mean rates of detection, identification, and avoidance by the three subject groups (**B**). Significant differences are indicated by asterisks (* = $p < 0.05$; ** = $p < 0.01$). EB = early blind; LB = late blind; and SC = sighted control.

Obstacle detection, identification, and avoidance: the EB group detected 54.9 ± 8.3% of the obstacles, whereas LB detected 45.2 ± 11.3% and SC 51.1 ± 14.1% of the obstacles. A two-way ANCOVA with correction for age and sex did not indicate any significant differences between groups in detection rate ($F(2,24) = 1.972$, $p = 0.428$, $\eta^2 = 0.070$). For obstacles that were successfully detected, the analyses did not find significant group differences for identification rate ($F(2,22) = 0.452$, $p = 0.642$, $\eta^2 = 0.039$). EB identified 88.4 ± 11.3% of the detected obstacles, LB 81.4 ± 10.6% and SC 81.6 ± 16.1%. Moreover, EB avoided 86.3 ± 11.8% of detected obstacles, LB 87.7 ± 7.2% and SC 68.3 ± 15.4%. Since the Shapiro–Wilk test signaled a departure from normality in avoidance scores, we applied the nonparametric Kruskal–Wallis test for this contrast. The analysis indicated a statistically significant difference between the groups ($H(2) = 6.844$, $p = 0.033$, $\eta^2 = 0.336$). Mann–Whitney post hoc tests with Bonferroni correction showed a significant advantage in obstacle avoidance for the EB ($p < 0.01$) and LB groups ($p < 0.01$) compared to SC, but no difference between EB and LB ($p > 0.05$) (Figure 3B). No effect of age nor sex was found in any of the analyses. For the LB group, there was no significant correlation with BDI in any of the measures ($p > 0.05$).

Detection according to the nature of the obstacles: The mean frequencies of cube detection were 55.7 ± 15.3% (EB), 43.0 ± 18.0% (LB), and 50.5 ± 21.8% (SC). The corresponding mean frequencies for postdetection were 32.1 ± 13.2% for EB, 25.9% ± 14.5% for LB and 33.3 ± 7.5% for SC. EB detected 35.1 ± 24.4% of steps, while LB and detected 16 ± 18.3% and 28.9 ± 19.8%, respectively. Doors were the most easily detected obstacles with mean performances of 97.2 ± 13.2% for EB, 95.0 ± 7.5% for LB, and 91.6 ± 10.6% for SC. The ANCOVA corrected for age and sex indicated a significant effect of type of obstacle $F(3, 94) = 94.141$, $p < 0.01$, $\eta^2 = 0.730$), but no group, age, or sex effects, nor for the interaction between groups and types of obstacles ($p > 0.05$). Post hoc tests with Bonferroni correction revealed that doors and cubes were significantly better detected than steps and posts ($p < 0.001$), but cubes were less well detected than doors ($p < 0.001$). No significant differences were found between detection of steps and posts ($p > 0.05$) (Figure 4A). *Collisions according to the nature of the obstacles:* The collisions rates were about 20% per type of obstacles in each group, except for steps, which had up to 80% collision rates. Indeed, the EB group collided with 68.3 ± 22.8% of steps, but only 16.9 ± 21.5% of cubes, 14.6 ± 14.1% of posts, and 6.3 ± 4.0% of door obstacles. For LB, results were similar with 80.2 ± 17.2% collisions with steps, but only 23.4 ± 16.9% with cubes, 18.6 ± 8.6% with posts, and 8.7 ± 7.9% with doors. In SC, there were 77.5 ± 22.7% collisions with steps, but only 22.8 ± 17.3% for cubes, 17.9 ± 10.9% for posts, and 34.0 ± 22.6% for doors. Much like the detection performances, the ANCOVA corrected for age and sex revealed a significant effect in the type of obstacles ($F(2, 94) = 79.811$, $p < 0.001$, $\eta^2 = 0.680$), but not in groups, age, or for the interaction between groups and type of obstacles ($p > 0.05$). Post hoc tests with Bonferroni correction indicated that collision with steps was significantly greater than for the other obstacles ($p < 0.001$). There was no significant difference in collision frequencies for cubes, posts, and doors. Moreover, the ANCOVA indicated a significant effect of sex in collision rates ($F(1, 94) = 6.132$, $p = 0.015$). However, the effect size ($\eta^2 = 0.017$) was very low, and the gender difference did not survive the post hoc Mann–Whitney test ($H(106) = 1140$, $p > 0.05$) (Figure 4B).

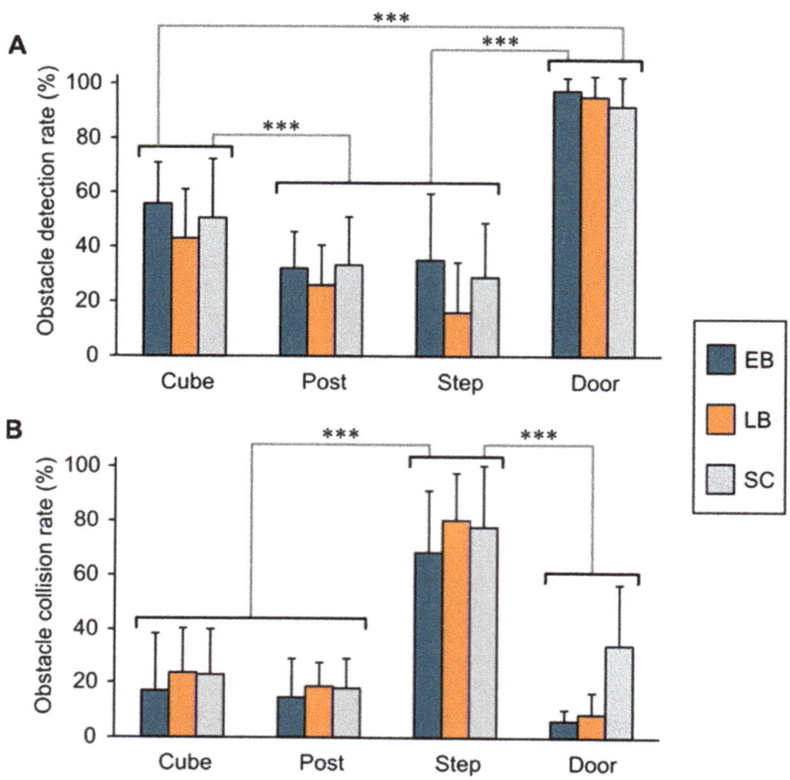

Figure 4. The average rates of obstacle detection (**A**) and collision (**B**). Significant differences are indicated by asterisks (*** = $p < 0.001$). EB = early blind; LB = late blind; and SC = sighted control.

4. Discussion

The goal of the present study was to investigate the reliability of the suggested EyeCane's downward sensor by testing a single-sensor EyeCane pointed toward the ground in the presence of obstacles that could induce tripping and falling in daily-life situations. Participants' performance in our obstacle course should serve as a reasonable indicator of the potential of the device as a primary mobility aid in various circumstances (i.e., indoor and outdoor travel). Furthermore, we wanted to investigate the differences in capacities and strategies between late- and early-blind individuals, to determine their implications for improving the design of the EyeCane and other mobility aids' design. We hypothesized that participants would have particular difficulty in detecting and avoiding the "step" obstacle, which would thus present a major limitation for the reliance on the EyeCane as a primary mobility aid. We also hypothesized that EB individuals, given their superior abilities with SSDs [13], would show better navigation capacities with the device than LB and SC participants.

The experiment was centered around three navigational tasks—obstacle detection, identification, and avoidance—which require distinct but interlinked processes to sustain safe navigation. In fact, these three processes all require an ability to attribute the device signal to specific objects in the external environment. This phenomenon, called *distal attribution*, is mandatory for navigation using devices such as ETAs and SSDs, and is mainly achieved during the training phase when participants calibrate their internal representation by touching obstacles [33]. In using the EyeCane to execute successfully our navigation task, the user must analyze the vibration intensity to extract information about the obstacle's presence and position and then analyze the vibration pattern generated by the scanning movement to extract the obstacle's identity (i.e., form and dimensions) and plan an avoidance path.

4.1. Influence of Visual Experience

Overall detection, avoidance, and identification performances were not statistically different between groups. This result might be surprising given the established literature on sensory substitution showing superior abilities of EB [34], but some studies on minimalist SSDs (Guidance-SSD; Sound of Vision) have also observed equivalent performances of EB, LB, and SC [24,35,36]. One might thus suppose that our findings of efficient performances for every group despite only brief training may be an indicator of the device's simplicity and ease of use [37]. However, both blind groups were significantly better at avoiding the obstacles they detected, while EB were significantly faster than LB and SC at doing the task. A possible explanation for this finding is that the EyeCane provides information on the relative distance between the user and the obstacle, which necessarily places the user at the center of his/her perceived space. This favors the use of an egocentric (body centered) spatial representation, which is known to predominate in EB, whereas LB and SC individuals are more used to working with allocentric (object centered) strategies [38]. Indeed, a normative visuocentric development favors spatial navigation behavior toward the use of an allocentric frame of reference [39]. We cannot exclude another possible explanation for the faster task performance of EB, namely the use of passive echolocation. Indeed, in O&M training, blind individuals are taught to use environmental sounds to obtain spatial information and detect objects [30]. Although we attempted to control for active echolocation by administering the test in a relatively nonechoing sonic and silent environment, we did not ask participants to wear earplugs. In addition, our blindfolding of sighted participants certainly placed them at a disadvantage, eliminating their usual visual inputs impairs their spatial abilities and postural control, even while using an SSD [6]. Therefore, these factors could have influenced the observed group differences in transit times for the obstacle course.

How the lack of visual experience developmentally affects the organization of the visual system and spatial representation could also be a factor in the present findings. Indeed, when deprived of vision early in life, the brain undergoes massive structural and functional reorganization that allows the visual cortex and its associative areas to recruit the other senses, often resulting in superior skills in these modalities (mainly touch and audition) [34]. Superior tactile acuity due to the rerouting of tactile input to the visual cortex in EB [40–42] could have facilitated their perception of vibration changes from the device, thus resulting in greater ease in obstacle detection and avoidance, leading to faster crossing times compared to the LB group. Due to these plastic phenomena, we expected EB to be more efficient in detecting the "step" obstacle and its faint vibration changes. However, their performance of this task did not differ significantly from that of the other groups, perhaps due to limitations in subject recruitment and the number of trials. Furthermore, several studies have shown that blind individuals can use the same structures of the navigational network that are normally involved in visually guided navigation (i.e., hippocampal formation, parahippocampal gyrus, posterior parietal cortex, and occipital areas) [10,43–45]. This circuit of brain regions seems to be vision dependent in sighted individuals, since it is not recruited when they navigate while blindfolded [10]. Moreover, a recent study with a tactile SSD reported that the EB recruit a sensorimotor circuit (e.g., inferior parietal cortex and areas 3a and 4p) when learning obstacle detection, while the sighted use mainly the medial temporal lobe (e.g., hippocampus and entorhinal cortex) [44]. Such crossmodal reorganization can also be present in the LB brain, albeit to a lesser scale and enhanced tactile acuity has also been observed in such individuals [46]. However, these processes are training dependent and, given that the cortex of LB is initially shaped by vision, the functional reallocation may not occur to the same extent as seen in the EB [47,48]. Although we found no significant effect of BDI on any of the results, this could reflect the considerable heterogeneity of the groups with respect to blindness onset, duration, and cause.

4.2. The EyeCane as a Primary Mobility Aid

The results obtained in this study are in line with previous studies on the EyeCane [23–25]: our participants were able to detect and avoid a variety of obstacles with good reliability (overall detection and avoidance performance) with little training (15 min). However, this study differed in the way subjects were instructed to scan with the device. Our participants scanned toward the ground with a technique resembling their daily use of the white cane, with the goal of detecting small obstacles or denivelations such as curbs on which they might trip and fall. While this technique allowed the efficient detection of obstacles such as walls and door frames, thin posts, and knee-high obstacles and was successful in sustaining good identification and avoidance performances for such obstacles, participants were less successful in detecting and avoiding the 0.15 m high "step" obstacle. In line with our hypothesis, this can be explained by the very faint vibration amplitude difference between the ground and the top of the "step" obstacle, which is difficult to perceive. A previous study with the EyeCane also using differently sized obstacles likewise demonstrated that bigger obstacles occupying the entire width of the hallway (similar to our "door frame" obstacle) were easier to detect than smaller and lower obstacles (like our "cube", "step", and "post") [26].

Present evidence shows that the EyeCane is an effective tool for detecting high obstacles by providing simple and relevant feedback to the user during mobility, even when the device is pointed toward the ground. However, this instrumentation seems insufficient for properly detecting those obstacles located at ground level ("step" obstacle), which is a mandatory function for safe navigation in both indoor (i.e., stairs) and outdoor (i.e., curbs and potholes) environments. Therefore, our results suggest that the EyeCane in its present state does not suffice as a primary mobility aid but rather can serve as a secondary aid used as an attachment to the white cane. The tactile signal given by the device in addition to the information obtained with the white cane would greatly augment the spatial information the users can acquire, increase their understanding of their surroundings, and thus increase their safety and independence. Indeed, our results show that the EyeCane would likely complement the white cane as an added protection against obstacles higher than the waist but also as a tool to identify (i.e., size, shape, etc.) these looming obstacles, which is crucial information to devise an avoidance strategy.

However, we note that this study took place within an experimental period of only three hours. We can thus hypothesize that with greater training and usage time, participants would likely get better at detecting and avoiding ground level obstacles. Nonetheless, given the absence of physical contact with the ground (normally provided by the tip of the white cane), it is unlikely that a long-term user of the EyeCane alone would reach the proficiency of a white cane user. Indeed, the sensors used in the device have certain limitations with regard to their sensitivity to environmental conditions and accuracy.

4.3. Implications for ETA Design

This study tested the reliability of a single-sensor EyeCane pointed toward the ground. Since our experimental design allowed to test its reliability to detect ground obstacles as well as higher obstacles, our results show that this device as a sole aid would not be sufficient to support safe navigation. Here, we list three significant limitations not only for the single-sensor EyeCane but also for the suggested three-sensors EyeCane (Figure 1D) [26], with general implications for the design of ETAs:

(1) *Loss of physical contact with the ground*. The white cane contact assures a high level of reliability in detecting ground obstacles and drop-offs such as curbs [49], and its replacement with the present downward sensor leads to less confidence in detecting drop-offs and ground obstacles.

(2) *Added complexity*. While the EyeCane's main advantage has always been its simplicity, ease of operation, and intuitive feedback, adding multiple sensors would add complexity and significantly increase the required training time. Indeed, a three-sensor design would necessitate feedback of a different nature (i.e., different modalities or frequency coding) for each sensor, as they each provide different spatial information that must be discriminated by the user.

(3) *User identification*. As the white cane allows pedestrians and drivers to identify the blind user and to assure his/her safety [3], a three-sensor EyeCane might lead to the loss of user identification and impede safety.

Despite these design limitations, the EyeCane (with a single infrared sensor) has proved its reliability in obstacle detection and avoidance in multiple studies [23,26,50]. With its advantages with respect to ease-of-use, simple feedback, light weight, and long battery life [23], the EyeCane may prove to be most useful as an attachment to white cane to improve the individual's safety, and reduce the risk of head injury [6]. In fact, all primary ETAs incorporate sensors attached to a physical cane (i.e., ultracane and WeWalk cane) [51,52], since blind individuals habitually rely heavily on the white cane's physical contact with the environment, as shown in previous studies on mobility aids [20,36]. However, these devices are often clumsy and expensive [53]. Thus, the low cost and ergonomic light weight, design of the EyeCane presents it as an affordable alternative for those who cannot afford an "all-in-one" electronic cane.

Further work should investigate the reliability and user acceptance [54] of a slim, ergonomic EyeCane fitted to the grip of a modified white cane's grip and equipped with an upward sensor. According to our findings, such an attachment would result in better obstacle identification and would substantially increase safety and route planning in the face of an environment encumbered by obstacles. We suppose that such a hybrid design would have great potential for adoption in the blind and visually impaired community worldwide.

5. Conclusions

This study aimed to investigate the EyeCane's potential as a primary mobility aid to avoid obstacles and diminish the risk of injuries and falling. Our results showed that the EyeCane efficiently enabled the users to detect, identify, and avoid large and high obstacles but failed to provide efficient coverage for obstacles at ground level. Indeed, removing the white cane and its physical contact with the ground leads to inconsistent detection of ground obstacles (i.e., steps, curbs, and holes) and higher risks of tripping and falling. Thus, the EyeCane is a potentially beneficial and low-cost attachment to the white cane that can significantly improve the individual's safety during mobility by providing coverage to obstacles above the waist.

Author Contributions: Conceptualization: M.B., S.P., I.D., R.K., and M.P.; methodology: M.B., S.P., and I.D.; validation, M.P. and R.K.; formal analysis, M.B. and S.P.; investigation, M.B. and S.P.; resources, D.R.C.; writing—original draft preparation, M.B. and S.P.; writing—review and editing, I.D., R.K., D.R.C., and M.P.; supervision, R.K. and M.P. All authors have read and agreed to the published version of the manuscript.

Funding: This research was funded by the Harland Sanders Chair in Vision Science (Université de Montréal, to M.P.) and the Canadian Institute of Health Research (PJT-9175018 to M.P.). M.B. was supported by Vision Health Research Network (VHRN, Quebec, Canada).

Institutional Review Board Statement: The study was conducted according to the guidelines of the Declaration of Helsinki and approved by the Comité d'éthique de la recherche clinique of Université de Montréal (CERC-19-097-P, 05-02-2020).

Informed Consent Statement: Informed consent was obtained from all subjects involved in the study.

Data Availability Statement: Data are available at ismael.djerourou@umontreal.ca.

Acknowledgments: The authors wish to thank Tomer Behor from RenewSenses Inc. for developing the EyeCane, Amir Amedi from the Baruch Ivcher Institute for Brain, Cognition, & Technology, Israël for making the instrument available, Paul Cumming from Inglewood Biomedical Editing (608 Millwood Rd, Toronto, ONT Canada) for excellent revision of the text and all the participants.

Conflicts of Interest: The authors declare no conflict of interest. The funders had no role in the design of the study, in the collection, analyses, or interpretation of data, in the writing of the manuscript, or in the decision to publish the results.

References

1. Schinazi, V.R.; Thrash, T.; Chebat, D.R. Spatial navigation by congenitally blind individuals. *Wires Cogn. Sci.* **2016**, *7*, 37–58. [CrossRef]
2. Ptito, M.; Bleau, M.; Djerourou, I.; Paré, S.; Schneider, F.C.; Chebat, D.-R. Brain-Machine Interfaces to Assist the Blind. *Front. Hum. Neurosci.* **2021**, *15*, 638887. [CrossRef]
3. Suterko, S. Long cane training: Its advantages and problems. In *Proceedings of the Conference for Mobility Trainers and Technologists*; Massachusetts Institute of Technology: Cambridge, MA, USA, 1967; pp. 13–18.
4. Manduchi, R.; Kurniawan, S. Mobility-related accidents experienced by people with visual impairment. *AER J. Res. Pract. Vis. Impair. Blind.* **2011**, *4*, 44–54.
5. Beggs, W. Coping, adjustment, and mobility-related feelings of newly visually impaired young adults. *J. Vis. Impair. Blind.* **1992**, *86*, 136–140. [CrossRef]
6. Chebat, D.-R.; Harrar, V.; Kupers, R.; Maidenbaum, S.; Amedi, A.; Ptito, M. Sensory substitution and the neural correlates of navigation in blindness. In *Mobility of Visually Impaired People*; Springer: Berlin/Heidelberg, Germany, 2018; pp. 167–200.
7. Kaczmarek, K.A. The tongue display unit (TDU) for electrotactile spatiotemporal pattern presentation. *Sci. Iran.* **2011**, *18*, 1476–1485. [CrossRef] [PubMed]
8. Ptito, M.; Matteau, I.; Zhi Wang, A.; Paulson, O.B.; Siebner, H.R.; Kupers, R. Crossmodal Recruitment of the Ventral Visual Stream in Congenital Blindness. *Neural Plast.* **2012**, *2012*, 304045. [CrossRef] [PubMed]
9. Matteau, I.; Kupers, R.; Ricciardi, E.; Pietrini, P.; Ptito, M. Beyond visual, aural and haptic movement perception: hMT+ is activated by electrotactile motion stimulation of the tongue in sighted and in congenitally blind individuals. *Brain Res. Bull.* **2010**, *82*, 264–270. [CrossRef] [PubMed]
10. Kupers, R.; Chebat, D.R.; Madsen, K.H.; Paulson, O.B.; Ptito, M. Neural correlates of virtual route recognition in congenital blindness. *Proc. Natl. Acad. Sci. USA* **2010**, *107*, 12716–12721. [CrossRef]
11. Pamir, Z.; Canoluk, M.U.; Jung, J.-H.; Peli, E. Poor resolution at the back of the tongue is the bottleneck for spatial pattern recognition. *Sci. Rep.* **2020**, *10*, 1–13. [CrossRef] [PubMed]
12. Richardson, M.L.; Lloyd-Esenkaya, T.; Petrini, K.; Proulx, M.J. Reading with the Tongue: Individual Differences Affect the Perception of Ambiguous Stimuli with the BrainPort. In Proceedings of the 2020 CHI Conference on Human Factors in Computing Systems, Honolulu, HI, USA, 25–30 April 2020; pp. 1–10.
13. Chebat, D.R.; Schneider, F.C.; Kupers, R.; Ptito, M. Navigation with a sensory substitution device in congenitally blind individuals. *Neuroreport* **2011**, *22*, 342–347. [CrossRef] [PubMed]
14. Elli, G.V.; Benetti, S.; Collignon, O. Is there a future for sensory substitution outside academic laboratories? *Multisens. Res.* **2014**, *27*, 271–291. [CrossRef]
15. Pissaloux, E.; Velázquez, R. On spatial cognition and mobility strategies. In *Mobility of Visually Impaired People*; Springer: Berlin/Heidelberg, Germany, 2018; pp. 137–166.
16. Islam, M.M.; Sadi, M.S.; Zamli, K.Z.; Ahmed, M.M. Developing walking assistants for visually impaired people: A review. *IEEE Sens. J.* **2019**, *19*, 2814–2828. [CrossRef]
17. Cardillo, E.; Caddemi, A. Insight on electronic travel aids for visually impaired people: A review on the electromagnetic technology. *Electronics* **2019**, *8*, 1281. [CrossRef]
18. Mattia, V.D.; Manfredi, G.; Leo, A.D.; Russo, P.; Scalise, L.; Cerri, G.; Caddemi, A.; Cardillo, E. A feasibility study of a compact radar system for autonomous walking of blind people. In Proceedings of the 2016 IEEE 2nd International Forum on Research and Technologies for Society and Industry Leveraging a Better Tomorrow (RTSI), Bologna, Italy, 7–9 September 2016; pp. 1–5.
19. Islam, M.M.; Sadi, M.S.; Bräunl, T. Automated Walking Guide to Enhance the Mobility of Visually Impaired People. *IEEE Trans. Med. Robot. Bionics* **2020**, *2*, 485–496. [CrossRef]
20. Hersh, M.; Johnson, M.A. *Assistive Technology for Visually Impaired and Blind People*; Springer Science & Business Media: Berlin/Heidelberg, Germany, 2010.
21. Cardillo, E.; Mattia, V.D.; Manfredi, G.; Russo, P.; Leo, A.D.; Caddemi, A.; Cerri, G. An Electromagnetic Sensor Prototype to Assist Visually Impaired and Blind People in Autonomous Walking. *IEEE Sens. J.* **2018**, *18*, 2568–2576. [CrossRef]
22. Maidenbaum, S.; Abboud, S.; Amedi, A. Sensory substitution: Closing the gap between basic research and widespread practical visual rehabilitation. *Neurosci. Biobehav. Rev.* **2014**, *41*, 3–15. [CrossRef]
23. Maidenbaum, S.; Hanassy, S.; Abboud, S.; Buchs, G.; Chebat, D.R.; Levy-Tzedek, S.; Amedi, A. The "EyeCane", a new electronic travel aid for the blind: Technology, behavior & swift learning. *Restor. Neurol. Neurosci.* **2014**, *32*, 813–824. [CrossRef] [PubMed]

24. Chebat, D.R.; Maidenbaum, S.; Amedi, A. Navigation using sensory substitution in real and virtual mazes. *PLoS ONE* **2015**, *10*, e126307. [CrossRef]
25. Maidenbaum, S.; Levy-Tzedek, S.; Chebat, D.-R.; Amedi, A. Increasing Accessibility to the Blind of Virtual Environments, Using a Virtual Mobility Aid Based On the "EyeCane": Feasibility Study. *PLoS ONE* **2013**, *8*, e72555. [CrossRef]
26. Buchs, G.; Simon, N.; Maidenbaum, S.; Amedi, A. Waist-up protection for blind individuals using the EyeCane as a primary and secondary mobility aid. *Restor. Neurol. Neurosci.* **2017**, *35*, 225–235. [CrossRef]
27. Meaidi, A.; Jennum, P.; Ptito, M.; Kupers, R. The sensory construction of dreams and nightmare frequency in congenitally blind and late blind individuals. *Sleep Med.* **2014**, *15*, 586–595. [CrossRef]
28. Innet, S.; Ritnoom, N. An application of infrared sensors for electronic white stick. In Proceedings of the 2008 International Symposium on Intelligent Signal Processing and Communications Systems, Bangkok, Thailand, 8–11 February 2009; pp. 1–4.
29. Chebat, D.-R.; Maidenbaum, S.; Amedi, A. The transfer of non-visual spatial knowledge between real and virtual mazes via sensory substitution. In Proceedings of the 2017 International Conference on Virtual Rehabilitation (ICVR), Montreal, QC, Canada, 19–22 June 2017; pp. 1–7.
30. LaGrow, S.J. Improving Perception for Orientation and Mobility. In *Foundations of Orientation and Mobility*, 3rd ed.; Wiener, W.R., Welsh, R.L., Blasch, B.B., Eds.; American Foundation for the Blind Press: New York, NY, USA, 2010; Volume 2, pp. 3–26.
31. La Grow, S.J.; Long, R. *Orientation and Mobility: Techniques for Independence*; Association for Education and Rehabilitation of the Blind and Visually Impaired: Alexandria, VA, USA, 2011.
32. JASP Team. *JASP (Version 0.14.1) [Computer Software]*; JASP Team: Amsterdam, The Netherlands, 2020.
33. Hartcher-O'Brien, J.; Auvray, M. The Process of Distal Attribution Illuminated Through Studies of Sensory Substitution. *Multisens. Res.* **2014**, *27*, 421–441. [CrossRef]
34. Kupers, R.; Ptito, M. Compensatory plasticity and cross-modal reorganization following early visual deprivation. *Neurosci. Biobehav. Rev.* **2014**, *41*, 36–52. [CrossRef]
35. Paré, S.; Bleau, M.; Djerourou, I.; Malotaux, V.; Kupers, R.; Ptito, M. Spatial navigation with horizontally spatialized sounds in early and late blind individuals. *PLoS ONE* **2021**, *16*, e0247448. [CrossRef]
36. Hoffmann, R.; Spagnol, S.; Kristjánsson, Á.; Unnthorsson, R. Evaluation of an audio-haptic sensory substitution device for enhancing spatial awareness for the visually impaired. *Optom. Vis. Sci.* **2018**, *95*, 757. [CrossRef]
37. Stronks, H.C.; Nau, A.C.; Ibbotson, M.R.; Barnes, N. The role of visual deprivation and experience on the performance of sensory substitution devices. *Brain Res.* **2015**, *1624*, 140–152. [CrossRef]
38. Iachini, T.; Ruggiero, G.; Ruotolo, F. Does blindness affect egocentric and allocentric frames of reference in small and large scale spaces? *Behav. Brain Res.* **2014**, *273*, 73–81. [CrossRef]
39. Pasqualotto, A.; Spiller, M.J.; Jansari, A.S.; Proulx, M.J. Visual experience facilitates allocentric spatial representation. *Behav. Brain Res.* **2013**, *236*, 175–179. [CrossRef]
40. Ptito, M.; Fumal, A.; de Noordhout, A.M.; Schoenen, J.; Gjedde, A.; Kupers, R. TMS of the occipital cortex induces tactile sensations in the fingers of blind Braille readers. *Exp. Brain Res.* **2008**, *184*, 193–200. [CrossRef]
41. Goldreich, D.; Kanics, I.M. Performance of blind and sighted humans on a tactile grating detection task. *Percept. Psychophys.* **2006**, *68*, 1363–1371. [CrossRef]
42. Stevens, J.C.; Foulke, E.; Patterson, M.Q. Tactile acuity, aging, and braille reading in long-term blindness. *J. Exp. Psychol. Appl.* **1996**, *2*, 91–106. [CrossRef]
43. Gagnon, L.; Schneider, F.C.; Siebner, H.R.; Paulson, O.B.; Kupers, R.; Ptito, M. Activation of the hippocampal complex during tactile maze solving in congenitally blind subjects. *Neuropsychologia* **2012**, *50*, 1663–1671. [CrossRef]
44. Chebat, D.R.; Schneider, F.C.; Ptito, M. Neural Networks Mediating Perceptual Learning in Congenital Blindness. *Sci. Rep.* **2020**, *10*, 495. [CrossRef]
45. Chan, C.C.H.; Wong, A.W.K.; Ting, K.-H.; Whitfield-Gabrieli, S.; He, J.; Lee, T.M.C. Cross auditory-spatial learning in early-blind individuals. *Hum. Brain Mapp.* **2012**, *33*, 2714–2727. [CrossRef]
46. Reislev, N.H.; Dyrby, T.B.; Siebner, H.R.; Lundell, H.; Ptito, M.; Kupers, R. Thalamocortical Connectivity and Microstructural Changes in Congenital and Late Blindness. *Neural Plast.* **2017**, *2017*, 9807512. [CrossRef]
47. Sadato, N.; Pascual-Leone, A.; Grafman, J.; Ibañez, V.; Deiber, M.-P.; Dold, G.; Hallett, M. Activation of the primary visual cortex by Braille reading in blind subjects. *Nature* **1996**, *380*, 526–528. [CrossRef] [PubMed]
48. Ptito, M.; Moesgaard, S.M.; Gjedde, A.; Kupers, R. Cross-modal plasticity revealed by electrotactile stimulation of the tongue in the congenitally blind. *Brain* **2005**, *128*, 606–614. [CrossRef] [PubMed]
49. Blasch, B.; LaGrow, S.J.; De l'Aune, W. Three aspects of coverage provided by the long cane: Object, surface, and foot-placement preview. *J. Vis. Impair. Blind.* **1996**, *90*, 295–301. [CrossRef]
50. Buchs, G.; Maidenbaum, S.; Amedi, A. Obstacle identification and avoidance using the 'EyeCane': A tactile sensory substitution device for blind individuals. In Proceedings of the International Conference on Human Haptic Sensing and Touch Enabled Computer Applications, Versailles, France, 24–26 June 2014; pp. 96–103.
51. Hoyle, B.; Waters, D. Mobility AT: The Batcane (UltraCane). In *Assistive Technology for Visually Impaired and Blind People*; Hersh, M.A., Johnson, M.A., Eds.; Springer London: London, UK, 2008; pp. 209–229.
52. WeWALK. WeWALK Smart Cane. Available online: https://wewalk.io/en/ (accessed on 11 April 2021).

53. Li, K. *Electronic Travel Aids for Blind Guidance—An Industry Landscape Study*; ECS: Berkeley, CA, USA, 2015.
54. Phillips, B.; Zhao, H. Predictors of assistive technology abandonment. *Assist. Technol.* **1993**, *5*, 36–45. [CrossRef] [PubMed]

sensors

MDPI

Review

Spatial Knowledge via Auditory Information for Blind Individuals: Spatial Cognition Studies and the Use of Audio-VR

Amandine Afonso-Jaco [1,2,*] and Brian F. G. Katz [3,*]

1 Laboratoire Mémoire, Cerveau et Cognition, Université de Paris, 92100 Boulogne Billancourt, France
2 Laboratoire Développement Individu Processus Handicap Éducation, Université Lumière Lyon 2, 69500 Bron, France
3 Institut Jean Le Rond d'Alembert UMR 7190, Sorbonne Université, CNRS, 75005 Paris, France
* Correspondence: amandine.afonso.jaco@gmail.com (A.A.-J.); brian.katz@sorbonne-universite.fr (B.F.G.K.)

Abstract: Spatial cognition is a daily life ability, developed in order to be able to understand and interact with our environment. Even if all the senses are involved in mental representation of space elaboration, the lack of vision makes it more difficult, especially because of the importance of peripheral information in updating the relative positions of surrounding landmarks when one is moving. Spatial audio technology has long been used for studies of human perception, particularly in the area of auditory source localisation. The ability to reproduce individual sounds at desired positions, or complex spatial audio scenes, without the need to manipulate physical devices has provided researchers with many benefits. We present a review of several studies employing the power of spatial audio virtual reality for research in spatial cognition with blind individuals. These include studies investigating simple spatial configurations, architectural navigation, reaching to sounds, and sound design for improved acceptability. Prospects for future research, including those currently underway, are also discussed.

Keywords: spatial cognition; spatial audio; blind navigation; binaural audio

1. Introduction

Spatial cognition is a general term we use to refer to a set of skills that are crucial in our everyday lives, such as representing the space around us and updating that representation as we move, locating, grasping, or pointing to external objects, learning routes, understanding maps, orienting, etc. [1]. It involves two major components: a dynamic one, such as navigation, and a static one, such as memorising the different objects or landmarks localisation, and/or the topographic knowledge [2,3]. In mental representation of space elaboration, there is an important role for peripheral information in updating the relative positions of surrounding landmarks when one is moving [4]. Extracting reliable positional information from multiple objects while moving is likely more difficult without vision as sight allows one to easily locate two or more landmarks in a space at the same time and to build relative references between them by using the triangulation method [5]. In comparison, for a person who is blind, the amount of distal information provided by the environment is reduced and the nature of available landmarks is more difficult to process [6]. Blind people's perception relies on haptic (tactile-kinesthetic) perception, which requires active exploration of the environment by the individual [7,8]. It involves complex processes that must integrate cutaneous information with proprioceptive and motor information related to the exploration movements [9]. For haptic exploration, mono- or bi-manual, the number of parallel landmarks that can be observed is limited, requiring blind people to collect a sequence of landmarks and to place them all together in their mental map which is, without vision, cognitively more complex, inducing congenitally blind persons to use egocentric coding strategies. Blind people are nevertheless able to acquire spatial information using other non-visual modalities, as shown by their performance in certain spatial tasks, especially when it involves the locomotor system (e.g., [10]). A large number of studies suggest that early visual experience is not a pre-requisite for the acquisition of spatial concepts (for a review, [11–14]). Blind people are able to mentally generate and correctly manipulate objects and gather spatial information provided by other modalities to create metrically valid internal representations, even though an early experience of vision facilitates the generation and use of mental images [15–19].

What appears to be clear from current literature is that there is no definitive position regarding the influence of the lack of early vision in spatial capacities. When we are interested in spatial cognition, there are as many different aspects that are studied as there are potential variabilities between individuals, due to their time of exposure to visual information, reasons that cause a person to be blind, and influence of both behavioural and cerebral reorganisation related to the visual deficit. Therefore, it is appropriate to focus on some specifics of spatial cognition and to draw conclusions only on those. As an example, in spatial abilities, there is a lot of discrepancy between different studies. Some demonstrate an impairment of early blind person's capacities: e.g., ref. [20] asked early blind and blindfolded sighted participants to make both direction (pointing) and distance (ratio scaling) judgements. They were invited to judge distances between their home (where the test took place) and other locations (either having been linked via walking by the participants, or otherwise imagined) or between two of these locations, considering themselves being at one of them. Results showed that early blind participants were less accurate in making *direction* estimations, but both groups reached the same level of accuracy in making *distances* judgements. Concerning the results where the location was imagined (lacking the physical memory of being there), the early blind group was less accurate than the sighted one. In a task where participants had to evaluate distances, point to sound sources, and estimate their coordinates, ref. [21] also showed that early blind performed less well. These conclusions are in line with the idea that early visual experience would allow for a more accurate understanding of external spatial coordinate systems [22,23]. In other words, having benefited from an early visual experience contributes to one's ability in evaluating spatial relationships between distal spatial cues.

However, other results coming from the similar types of studies suggest that primary visual experiences do not influence spatial abilities. Some studies asked participants to estimate the distances between landmarks or objects that had not been linked by a pathway during the initial phase of learning [24,25]. No difference between early blind and blindfolded sighted participants were observed. In the same vein, refs. [26,27] demonstrated that early or congenital blindness has little or no effect on direction and distance estimation of spatial relationships among locations that were actually visited by participants or explored with their fingers on tabletop tasks. A study on sensory substitution in the context of route finding in a simple maze using a sonified/haptified cane providing distance to obstacle information (a mono-aural/non-spatialized distance Geiger counter metaphor [28]), showed comparable performance between blindfolded-sighted and visually impaired individuals [29], supporting the notion that "representation of space is amodal (i.e., modality-independent)". Repeating the experiment for the same mazes via both physical exploration and a highly simplified desktop navigation showed that while task completion times did not generally improve (spatial memory of the correct route), false turns and associated errors reduced [30].

Focusing on the ability of creating mental representations of space which preserve the topology and metric relations between the different landmarks, suggesting that mental representation correctly informs them about the structure of the space around them, the results regarding blind people are interesting.

In a mental scanning task (originally developed in [31,32]), after a visual, haptic, or visual and haptic learning of a spatial configuration on a tabletop containing five objects, sighted participants performed similarly between the three learning conditions [33]. In the case of the haptic learning condition, all the participant groups (congenitally blind and blindfolded sighted) obtained the result classically obtained in the literature, i.e., a linear relationship between the times of mental exploration and the distances to be mentally explored, and this, without any significant differences between scanning times between groups.

These results are consistent with [11], which employed a mental rotation task with congenitally blind and blindfolded sighted participants, presenting the same conclusions about the non-crucial role of the visual experience for preserving the topology of a spatial configuration.

Other studies, still using a tactile exploration in the learning phase, such as [34], showed the same pattern of results, i.e., the linear relationship between the distance to be mentally travelled and the time needed to do so, for both sighted and blind (from birth or later) participants. Nevertheless, in contrast to [33], blind participants took significantly longer time than blindfolded to perform the task. These results show that even in the case of permanent visual deprivation, blind people were successful in elaborating structured spatial representations, even if this can lead to longer cognitive processing times.

These results are also in line with the work of [35,36] who showed, by employing a task of mental comparison of distances (originally proposed in [32]), that the spatial mental representations elaborated by both the blind (from birth) and the late-blind (temporarily deprived of vision or not) preserved the metric relations, as well as the topological organisation of the different elements composing the initial environment to be learned. However, the time required for the blind participants (both native and late-blind) to solve the task was significantly longer than that required for the sighted. These results support the hypothesis that permanent visual deprivation has an influence on the processes of elaboration of mental representations of space, just as in [37]. These authors have shown in a spatial inference task that participants who were blind from birth and late-blind had similar performances, but that sighted people went significantly faster to solve these tasks. Ref. [38] also interpreted these results as evidence that congenitally blind participants in particular have a slower mental process for solving this type of tasks.

It is interesting to focus on this precise point. One way is to consider that the extra time needed does not reflect a less accurate spatial ability, but more the translation of a more difficult-to-access spatial information.

We can hypothesise that the modality with which an environment is learned influences the metrical properties that the participant infers about the spatial arrangement. In particular, if this modality is more preferred/familiar to blind participants than to sighted ones, such as the haptic modality, the results of [33] are easier to explain compared to those of [34–36]. The fact that blind people are slower in the task can be explained by the hypothesis that sighted people directly access a visual representation of the spatial layout and, therefore, only need to "look" at this representation to provide answers. In contrast, people without visual experience must translate the requested information into a more informative (haptic or locomotor) representation by mentally simulating their own movements (walking or "finger running" in a small environment). This "translation" explains why congenitally blind participants required more time to complete the same task.

In the absence of vision, locomotor experience is an alternative source of information for building mental representations of an environment. The onset of independent locomotion has been shown to be a pivotal event in the life of the human infant, triggering surprisingly far-reaching changes in a variety of psychological functions, including coordination of perception and motor skills, spatial cognition, memory, and social and emotional processes. Indeed, moving from one place to another reveals meaningful information about the environment in which one is walking, and all the more so, in the case of blindness [39].

Much work has been performed to investigate the ability of blind people to navigate complex environments without relying on visual information [10,27,40,41]. However, little was known about the nature and structure of the representations blind people use to underpin their navigational performance. Again, image scanning has been shown to be a reliable means of assessing the metric properties of mental images and providing information about the structure of blind people's mental representations in space.

Ref. [42] investigated several variations of a situation in which sighted participants learned about an environment by walking along paths connecting salient landmarks and then performing a mental scanning task. Results showed that scan time increased with scanned distance, suggesting that the mental representation of space based on locomotor exploration preserves information about the relative distances between different locations. In another experiment, they compared two learning conditions: one in which the path was visually inspected (without locomotion), and the other in which the path was physically walked without seeing it (blindfolded and guided by the experimenter). The scanning task again showed that both learning conditions resulted in the same typical *time* ∝ *distance* correlation, but that absolute scan times were shorter under visual than under locomotor learning conditions. This finding was consistent with the results of studies showing that vision and visual strategies improve the speed and accuracy of spatial performance [1,12,43]. The next step was then to propose the same type of experiment to blind participants in order to clarify whether their mental representations preserved the metric properties of the learned environment. One issue was to distinguish between congenitally blind and late-blind people, since locomotion is by far the most important means by which congenitally blind people learn large spaces.

One way to further investigate these questions is the use of the virtual audio 3D rendering, allowing researchers to construct complex scenes which would be unfeasible in real contexts. The ability to play back individual sounds at specific positions or to create complex spatial audio scenes without having to manipulate physical devices (e.g., silently moving speakers or even walls) offers many advantages. Thus, the use of spatial audio is no more only focused on the study of low-level processes, such as localisation, and is being used as a tool to study higher-level cognitive process.

We would be remiss if we did not mention the sense of *hearing*, and the widespread assumption that blind people have a distinct sense of hearing compared to the sighted population. Numerous studies have been conducted to investigate this claim which focus on a variety of aspects, from spatial precision to reaction time, neural plasticity, and brain activity, (for a review, see e.g., [44]). Although localisation, spectral analysis, and other basic tasks are generally considered of significant importance in understanding basic auditory perception and performance differences between sighted and blind individuals, these performance differences are inherently limited by the capabilities of the human auditory system. Rather, it is how this acoustic and auditory information is used, requiring higher level cognitive processing, where blind individuals can outperform sighted people. An example of where this is evident are navigation tasks. Ref. [45] conducted an experiment where participants walked at a constant distance from a simple straight barrier, being either a wall or a series of 2 m spaced poles, without making physical contact with the barrier. Finger snaps and footfall sounds were the only information. Compared to 14 blind sighted control participants, the 8 blind participants clearly outperformed the control group, some of whom actually considered the task impossible. Results showed that, overall, the blindfolded sighted subjects performance in the *wall* condition was comparable to blind participant performance in the *pole* condition.

With regards to navigation within an architectural space, ref. [46] investigated spatial navigation with sighted and blind children (aged 4.5–9 yrs). In a carpeted room, a tactile landmark was situated in the centre of each of the four walls. In a learning phase, blind or blindfolded participants were guided within the room to the set of landmarks, facilitating the creation of a spatial cognitive map, either with or without the presence of an auditory landmark, i.e., a single metronome ticking at the starting position. Participants then were invited to follow certain trajectories between the tactile landmarks, incorporating familiar paths from the learning phase, as well as novel paths. Results for sighted subjects showed improvements correlated with age, as well as for the auditory landmark condition. Considering only the new paths, all groups benefited from the auditory landmark. In the final *distance* error analysis, the sighted children performed better than the blind subjects in both conditions, with the blind subjects in the condition with the auditory landmark performing comparably to the blindfolded subjects without the auditory landmark. It should be noted that due to the protocol used, it was impossible to separate the effects coming from the auditory landmark or the learning effect.

One can remark the difference in outcomes of the two above cited studies, where, in [45], blind participant children out-performed sighted in auditory obstacle identification, while in [46], blind participants under-performed sighted in a mental map construction through real navigation. Of course, the tasks are rather different in the two, where the first addresses a skill where blind individuals train (unlike sighted individuals) while the second addresses a general skill (mental map construction) via various learning means. As the second study showed improvement for both groups with age, one can assume a potential lag in acquiring this skill by blind individuals, potentially linked to visual-centric notions employed in mental map development at a younger age, ill-suited or more difficult for blind individuals to acquire. This is similar results observed in the study detailed in Section 3 concerning scenario reconstruction after active locomotion.

We present in the remaining sections a review of several experiments conducted using virtual audio 3D rendering to enhance spatial cognition in blind participants, offering direct simulation of a blind individual's classical interaction with the world. Moving through it and gathering locomotor as haptic feedback and auditory ones. Here, some reports of previous studies conducted in collaboration between researchers from psychology and acoustics on the topic of spatial cognition shedding light on how blind people are offered a better way to investigate their mental spatial representation, their results appears to be quite similar to those of sighted individuals. Section 2 presents a brief introduction to the concepts employed in spatial audio or auditory virtual reality. Section 3 concerns the role of auditory information in assisting the creation of spatial knowledge. Section 4 investigates whether a blind person is able to collect meaningful acoustic information via an audio VR system designed to deliver a realistic 3D experience of the to be learned environment. The review then concludes with perspectives of research, following the same line of inquiry.

2. Brief Introduction to Auditory Virtual Reality

Binaural technology is the solution for sound spatialization that is the closest to natural hearing in the real world. Binaural reproduction attempts to mimic the entire set of acoustic cues involved in human localisation of sounds by reproducing the corresponding acoustic pressure at the entrance of the auditory canals via headphones. These two signals are necessarily a complete and sufficient representation of the sound scene, as they represent all the information exploited by the auditory system to identify the 3D position of a sound source. As such, binaural reproduction of spatial information is fundamentally based on the production (via recording or synthesis) of the collection of localisation cues, comprising the ITD (Interaural Time Difference), the ILD (Interaural Level Difference), and monaural spectral cues [47]. Taken together, the effects of these various cues are collected in the so-called Head-Related Transfer Function (HRTF), characterising the spectro-temporal filtering of an incident sound due to the morphology of the listener's head, torso, and pinnae. The ILD and ITD, varying as a function of source position, are principally determined by the individual's head size and shape, as well as the position of the ears relative to the head centre.

For more details on this topic, please refer to the following texts [47–50]. This method is at the foundation of much of today's virtual reality systems and has been used in the development of navigation guidance systems for the blind (e.g., [51]).

The general motivation for room acoustic modelling has been to enable the construction of acoustically better environments [52]. The most common method is geometrical acoustic, generally employing the technique of *ray tracing*, which models the propagation of sound through an analogy of light rays. In tracing the propagation of thousands, even millions, of rays and their interaction with the surfaces in a complex geometry, the room acoustic response can be estimated. Through such simulations, and subsequent "auralization" of the simulated room acoustic, it is then possible to render *audible* the acoustics of a computer simulated space [53,54]. While employed in the acoustic design of buildings [55,56], the auralization of simulated room acoustics is also at the centre of high quality virtual reality simulations which have been used, in, e.g., multimodal perception research [57,58] and historical reconstructions [59–61]. These techniques of binaural audio and room acoustic auralization are employed in several of the studies presented here.

3. Comparing Small-Scale Spatial Configuration and Locomotor-Scale Environment on Blind'S Mental Representation Properties

We examine in this set of studies to what extent the size of the environment, as well as the use of an active exploration of this environment by immersion in a virtual world in 3D audio, during the learning phase could have an influence on the metric properties of mental representations. Previous studies examined mental scanning, as well as performance on distance comparison tasks after verbal or haptic learning in blind (congenital or late) and sighted (blindfolded or with unimpaired vision) subjects [15]. The main conclusion was that blind people, like sighted people, were able to generate an accurate mental representation of an environment, which they obtained either by verbal description or by haptic exploration of a small-scale configuration. However, blind people needed significantly more time.

The metric properties of proximal spaces experienced without any motion component might be particularly difficult to encode for congenitally blind people. For them, the optimal conditions for constructing spatial representations are likely to depend on body involvement [27,40,42]. If this is true, it is reasonable to assume that congenitally blind people perform better on tasks in which the acquisition of spatial knowledge is based on locomotion.

The working hypothesis was that mental representations of spatial configurations may be based on different strategies. A sighted person might use mental representations that are iconic in nature, whereas blind persons might better remember sensorimotor contingencies. Ref. [62] have suggested that visual images, even for sighted people, can be representations based on information gathered through a number of different sensory modalities. Ref. [15] elaborated a new experimental context that further opens up the question of how blind people represent space-namely, the role of auditory information in supporting the creation of spatial knowledge [21,63].

The aim was to evaluate the effect of visual experience on mental spatial representation. Two learning conditions were contrasted, one designed to elicit a mental map of an iconic nature, while the other produced a mental one based on perception/action couplings. The comparison was intended to help determine which learning modality produced the most accurate representation of a spatial configuration, as well as to gather information about the perception by blind and sighted people of a world interacting with a three-dimensional audio environment. A large-scale immersive virtual audio environment (absent of visual feedback) with binaural audio and dynamic position and orientation tracking was developed where participants were able to explore and interact with local virtual sound objects (although the boundaries of the physical and virtual worlds aligned, the room acoustics of the space was not represented in the virtual simulation) (see Figure 1). The goal was to investigate mental spatial representations as a function of visual experience. Comparing two learning conditions, the first conceived to elicit a mental map of an iconic nature, the second intended to produced mental maps based on perception/action couplings. This comparison examined which learning modality produced the most accurate spatial configuration representation, in addition to gathering information about the perception by blind and sighted individuals within an interactive three-dimensional audio environment.

Participants performed several tasks, including manual reconstruction of the sound scene after a learning phase (see Figure 2) and tasks involving the proposal of different virtual scenes with minor and major metrical changes, followed by questions and the opportunity to correct the scene. Finally, mental scanning and mental distance comparison tasks were proposed [15,36,44].

(a) Physical and virtual scene of the experiment (shown on a meter scale). (b) Screenshots of the (left) subjective and (right) overview displays of the virtual scene, for monitoring by the operator.

Figure 1. Virtual reality scenario with six sound sources constructed for the mental representation study [64].

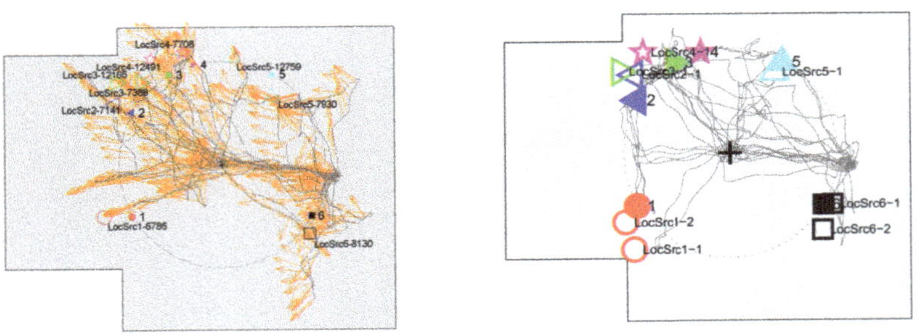

(a) Example of a schematic representation (visualisation of an experimental "log") of the trajectories of a participant, including head orientation, in the recreated virtual scene: *solid* symbols on the periphery of the circle (dotted line) represent the correct location of each sound source. *hollow* symbols represent the location chosen by the participant to position the sound source.

(b) Alternate reconstruction example highlighting repositioning refinement for achieving final positions for analysis.

Figure 2. Examples of scenario reconstruction for the mental representation study [44,64].

3.1. Protocol

Many previous experiments in the literature employing 3D audio for sound localisation studies or in the development of audio interfaces for blind individuals rely on specialised platforms for spatial rendering of audio with little or no graphical element. Whether using real sound sources [65] or a virtual environment [66], position/orientation user tracking is needed and some concept of a geometrical scene model must be updated in real time. Although the basic components were also present in this study (tracking and scene representation), the selected approach to virtual audio modelling was novel at the time, relying on real-time scene graph incorporating multimedia 3D effects, behaviour modelling, and interaction (VirChor [67]). The rationale was due to the complexity of the experimental setup, requiring the experimenter to monitor the position of both participants and active audio sources. In addition, the protocol's complexity and its progressive refinement benefited from an open scripting language that could be easily modified by non-programmers. The spatial sound and graphic rendering system architecture and software designs, including elements for dynamically modifying the behaviour of the user interface in response to participant and experimenter inputs are detailed in [64].

The paradigm of scanning and mental comparison of distances was used, with the goal of contrasting verbal and locomotor experiences as sources of learning. This meant that participants had to construct mental maps of more extended full-scale spatial environments, rather than the small configurations previously used in first experiments [35,36]. To this end, we designed an experimental situation suitable for studying blind people's abilities to construct spatial representations of an environment filled with sounds (see Figure 2). The aim was to evaluate the structural properties of the spatial representations acquired by blind and blindfolded sighted participants in two situations, namely, listening to a verbal description of the locations of different sound sources and physically moving within the environment to spatially locate and position each source.

The procedure consisted of immersing participants, people who are congenitally or late blind and blindfolded sighted participants in a real environment large enough to allow for locomotion. Participants were provided with a virtual audio sound scene, i.e., the virtual experience of a spatial auditory scene was created that consisted of an organised set of natural sources distributed in space. Using the VR platform, the rendered auditory scenes provided participants with the opportunity to interact directly with the different sound sources (approach them, move away from them, walk around them, etc., see Figure 1). The VR platform included a tracking system that captured and recorded participants' movements as they moved through the environment. The VR tracking system ensured that the virtual auditory scene remained stable in space despite shifts or head movements. Participants reported feeling as if they were surrounded by sound sources perceived at precise positions, creating a coherent spatial environment, as is usually the case in natural environments with real fixed sound sources. Among the advantages of the VR platform was the ability to control the exact geometric position of each sound, which could be changed dynamically without repositioning physical devices. After becoming familiar with the environment using one of two modes (the verbal and locomotor conditions), participants were tested with the scanning and mental comparison of distances paradigm.

3.2. Results

The mental scanning results showed that in the case of learning by verbal description, all participants, both blindfolded sighted, late blind or congenitally blind, showed a correlation between mental travel times and distances to be travelled, which was positive and significant. Thus, the greater the distance between two sound sources, the longer the associated travel time, as in the original reference studies [31,68].

In the *verbal* description learning condition, results suggested that in the case of immersion in a full-size 1:1 scale navigable environment, visual experience did not affect mental representations of the configuration, which preserves the metric relations maintained between the different sound sources. Nevertheless, it should be noted that individuals without early visual experience obtained lower correlation coefficients than late-blind and blindfolded sighted individuals.

In contrast, results obtained in the *active exploration* condition showed that for the congenitally blind group, the correlation between distances and mental travel times, although positive, did not reach the significance threshold. In view of the dispersion of the data points, a larger number of participants would seem to be necessary to resolve this question.

Furthermore, results for mental scanning times showed a clear difference in the behaviour for the blind group following the learning of either a small size configuration or a large size environment. Although the classical positive correlation was observed between distances and mental travel times following immersion in the large environment, no time/distance correlation was observed for the small size configuration, for either verbal description or after tactile exploration, reflecting a poor internal organisation of their mental representation.

These results suggest that, in the case of learning the spatial configuration of a locomotor space, early deprivation of visual experience did not have a specific impact. It seems however that early visual experience plays a dominant role in expertise regarding smaller differences, a hypothesis supported by results of the task of mental comparison of distance on small configurations. It was shown that the blind from birth group made significantly more errors in the case of small differences in distances than late blind and blindfolded sighted participants. Furthermore, whereas following learning of a small size configuration, congenitally blind participants took significantly longer than blindfolded sighted participants to compare distances, this was no longer the case after learning the spatial configuration in the immersive environment.

Results obtained with the mental comparison of distance task concluded that regardless of the learning modality of the environment in which they were immersed, all were able to correctly judge the length of two distances, and this independently of the visual experience they may or may not have had. The greater the difference between two distances, the greater the percentage of correct responses, and this for all groups of participants, as seen previously [68]. It was further observed that participants made more errors, and took more time to judge the differences, for small differences in distances, compared to medium or large differences. Response times also showed the same results as those obtained in the literature on smaller configurations, namely that small differences in distance generate longer response times, as compared to medium and large differences in distance [68].

These results suggested that the processes involved in the scanning and mental comparison of distances tasks, although both informing about the preservation of the metric qualities of mental representations, call upon different mechanisms that may or may not be affected, depending on the learning condition. One plausible explanation is that the congenitally blind relied exclusively on other strategies and a form of non-visual spatial imagery to form their mental representations. These strategies would be less effective than the visual imagery used by sighted groups, thus requiring a longer developmental process [69]. In contrast, the larger the initial environment, the less costly this difference in strategy would be, approaching the performance of sighted people in solving these tasks.

From these experiments, one can unequivocally conclude that early visual deprivation does not affect the ability of subjects to correctly represent the environment by preserving their structural isomorphism. In other words, in the case of early visual deprivation, no limits emerged regarding the adaptive capacities to allow the individual to represent environments correctly (i.e., preserving the topology and metric relations between elements). The differences, particularly in response times, suggested that early visual deprivation results in the use of more costly strategies than simply using a mental image, resulting in longer response times. It appears that the *size* of the environment, rather than the learning *modality*, was the determining factor. This alternative explanation deserves more systematic investigation, given the differences between small-scale and large-scale spatial abilities [70].

Regarding the reconstruction of the learned sound scene, the initial hypothesis was that learning through active exploration would provide an advantage for blind participants over learning through verbal description. A second hypothesis involved sighted participants, who were expected to benefit more from verbal description because they were better able to create a visual image of the scene and thus more accurately recreate the original configuration of the scene. The results suggest that active exploration of an environment improves learning of the *absolute positioning* of sound sources compared to verbal description. The same improvement was shown with respect to *radial* distance errors, but only in blindfolded participants. Results demonstrated that both blind participant groups underestimated circle size regardless of learning modality, with mean position error close to zero, and clearly benefited from learning with perception-action coupling.

These results are similar to [20], where blind participants were less accurate in making *direction* estimations, but all groups reached the same level of accuracy in making *distances* judgements. However, they are not in line with others, such as [71] in which blind subjects performed better in estimating the distance to real sound sources by only turning their heads and verbally reporting their position. It is clearly shown that active exploration of the environment improved the performance of blindfolded participants, both in terms of absolute position and size of the reconstructed configuration. It was also observed that subjects who were blind from birth made significantly more errors in angular positioning than late blind or blindfolded groups in both learning conditions. These data are consistent with the results of previous studies dealing with the processing of spatial information in classical real (not virtual) environments [72].

4. Wayfinding

Another significant interest in using spatial audio virtual reality is the ability to offer the opportunity to blind people to learn new unknown environments before a real navigation (i.e., going there). Navigation in an enclosed environment requires analysis of a variety of acoustic cues, a task that is well developed in many visually impaired people and for which sighted people rely almost exclusively on visual information. For blind people, creating cognitive maps for spaces such as homes or buildings can be a tedious process.

If one considers the context of a blind individual arriving at a new job site, they could be expect to rely on exploring the building after hours, or at a minimum, when occupancy is very low, to actively explore the architectural environment. The goal of such navigation is to gain some knowledge of spatial configurations and basic characteristics of the acoustic environment, while avoiding being disturbed by other people (including reverberation and echoes, footfall noises, etc.). Later, the individual becomes increasingly familiar with the daily sounds associated with different areas of the environment. The study presented in the following section poses the question of whether a blind individual could gather such acoustic information—critical to subsequent adaptation—using an audio VR system capable of presenting a realistic 3D rendering of the environment of interest. The valuable aspect of a VR system would be the possibility to perform the acquisition phase at one's own discretion, i.e., in a different place other than the actual environment, e.g., the user's home or in a public resource centre. In short, is it possible for a person to learn an architectural environment without being physically present? If so, such a system could prove useful for navigational preparation in new and unfamiliar environments.

A comparison of two types of learning was proposed: in situ real displacement versus active navigation in a virtual architecture. In these two conditions, participants were not allowed to use white cane or be accompanied by a guide dog, only acoustic information was available [73].

The study was designed to provide information about the spatial configuration of an enclosed environment through the use of interactive virtual acoustic models. Two interactive 3D acoustic models were created that simulated the two real environments, and an experiment was conducted comparing the mental representations elaborated after real and virtual navigation.

4.1. Protocol

Previous observations in the real-world navigation phase have shown that blind individuals make extensive use of self-generated sounds, such as finger snaps and footsteps to determine the position of an object (wall, door, table, etc.).

For that reason, simulation of these self-produced noises was included within the two 3D interactive acoustical models. Regarding the ability of the system to provide an accurate virtual auditory environment, an HRTF selection phase was performed by each individual so that an optimal/individualised binaural rendering could be presented (see [74] for more details on this procedure).

In the experimental condition, participants were given a joystick as a navigation control device and headphones equipped with a head-tracking device. Stepping sounds were automatically reproduced according to the participant's movement in the virtual environment and corresponded approximately to a 50 cm step. Navigation speed was continuously variable, ranging from 0.1–1 ms^{-1}, proportional to the angle applied to the joystick, i.e., the further it was pushed forward, the faster one advanced, and vice-versa. The finger snap was played as soon as the joystick button was pressed. As movement in the virtual space was limited to a single linear 'track', with the ability to only move forward or backwards (all other actions of the joystick were ignored), and head orientation being determined by the actual head orientation of the participant, very little learning time was required for familiarity.

The experimental evaluation consisted of comparing two types of navigation (real and virtual) along the two different corridor environments, with participants given the opportunity to walk back and forth along the path as many times as desired until they were confident enough that they understood the spatial configuration of the environment itself (see Figure 3). In both conditions, all participants stopped several times along the path and listened for elements of the environment. Active, self-generated sounds (their fingers snapping), along with footfall sounds, were reported to be extremely useful for understanding the configuration of specific parts of the environment.

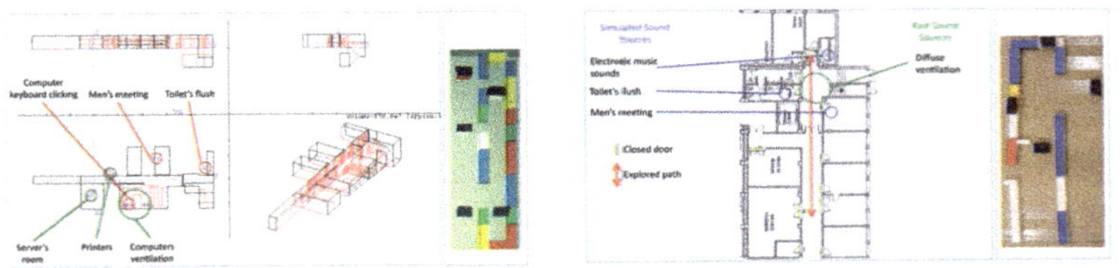

(a) Geometrical acoustic model of the first space including positions of sound sources and example block reconstruction following navigation.

(b) Plan of the second space scenario including positions of sound sources and example block reconstruction following navigation.

Figure 3. Virtual reality scenarios of the architectural test spaces and example block reconstructions for the architectural perception study [44].

To perform the mental representation assessment tasks, a number of sound sources were placed at specific locations in each environment, and both the sounds and their positions were the same in real and virtual navigation.

Participants were asked to try to mentally represent the configuration of the environment very accurately, taking into account as many of the numerous elements and relative positions as possible (changing floor coverings, openings, windows, doors, objects, obstacles, etc.). In the real condition, two congenitally blind and three late blind subjects (three women, two men) participated in the experiment, while in the virtual condition, four early blind and one late blind individual (three females, two males) explored the same two environments. In order to access the quality of the mental representation blind's participants elaborated, a mental comparison of distances task between the various sources in the real/virtual environments was employed.

4.2. Results

Due to the low number of participants, the results of this study could be described as having low statistical power, nevertheless, results were very informative about the question considered.

The results showed that even at a high level of performance for the real navigation condition, the symbolic distance effect was still confirmed. The probability of making a correct decision when two distances were mentally compared increased with the magnitude of the difference. A similar trend was found in the virtual navigation condition. These results illustrate that the variable real/virtual cannot be considered statistically relevant and that no significant differences were observed between the real and virtual navigation modalities for all three types of distance differences (small, medium, and large). This initial evaluation tended to confirm that both physical displacement (real navigation) and active virtual navigation using a joystick in a virtual architectural acoustic environment enabled blind subjects to create mental representations that preserved the topological and metric properties of the original environment.

Participants also used LEGO® blocks to reconstruct the representation they made from the environment, combined with verbal annotations, according to the mental representation they elaborated from their navigation (real or virtual), see Figure 3. Participants were asked to be as detailed as they could, giving comments concerning the walls, floors, different sound variations, etc.

The results suggest that exploration within an interactive virtual acoustic environment, such as the one developed here, was sufficiently realistic and well defined to provide appropriate information for spatial understanding of the architecture. An interactive virtual simulation could be precise enough so that information about the spatial configuration of the entire environment, not just the positions of sound sources, could be grasped by visually impaired individuals through auditory exploration alone. In short, interactive VR navigation performed well because it included both head tracking and controlled displacement. These results were of particular value because no significant difference was found between the real and virtual navigation conditions in the behavioural measurements. Joystick-controlled navigation enabled participants to build mental representations of indoor spatial structure that included both sound sources and realistic room acoustic simulations. This mental representation preserved the topological organisation and metric properties of the environment, as was the case with real navigation.

5. Discussion

The main objective of this paper was to shed light on how spatial audio technology is able to help to investigate spatial cognition, and in particular that of the blind persons, in a new manner. Many previous studies have been constructed to investigate the way primary visual experience can influence spatial cognition, elaborating more or less complex protocols, and leading to controversial results. Particularly in the field of spatial cognition, no one has concluded definitively about the influence of the lack of (early) vision on spatial capacities, likely because of the varied specific aspects that can be studied, including the potential variability between individuals. The major variable being blind congenitally or becoming blind later in life. Numerous studies indicate that early visual experience is not a prerequisite for the acquisition of spatial concepts (for a review, [11–14]), implying that blind people are able to mentally generate and correctly manipulate objects, gathering spatial information provided by other modalities, to create metrically valid internal representations, even though an early experience of vision facilitates the generation and use of mental images [15–19]. In the presented studies, we focused on a particular spatial ability of blind people, specifically, the ability of creating mental representations of space preserving the topology and metric relations between the different landmarks, suggesting that their mental representation of the configuration correctly informs them about the structure of the space around them, which is crucial in their every day life.

The principle classical tasks informing on this specificity of spatial cognition come from the mental scanning of distance task (originally proposed in [31,75]) and from the mental comparison of distance task (originally proposed by [32], adapted to blind subjects by [36]). After a haptic learning of the configuration, congenitally blind participants' data displayed the same linear relationship between scanning time and distance as those of (blindfolded) sighted people, the absolute scanning times were not significantly different between sighted and blind people [33], contrary to Kerr [34], who also reported a strong positive correlation between scanning times and the length of mental travel for both participants, but with significantly longer time for blind than for blindfolded persons. This finding suggested that the structure of mental representations of spatial configurations can be achieved despite the absence of sight, although the cost of generating and scanning these representations could be higher for blind people.

Refs. [35,36] showed that the topological organisation and metric relationship between the objects composing a spatial layout were preserved in the mental representations constructed by blind persons in a mental distance comparison task. However, the analysis also showed that the time for mental processing of distances was longer in blind individuals (from birth and later) than in blindfolded individuals, suggesting that definitive visual deprivation affects the way mental representations are processed.

One way to consider the extra time needed by the blind person is to imagine that it is the reflection of a slower mental process of the information, or a more difficult access to spatial information, and not a reflection of a less accurate spatial ability. These differences could be due to the modality under which an environment is learned, and if it is or not a preferred blind modality. This could explain why sometimes blind performances are better or alternatively less accurate than those of blindfolded sighted ones.

In the absence of vision, locomotor experience is an alternative source of information to elaborate mental representations of an environment, but little was known about the nature and structure of the mental representations blind people used to underpin their navigational performance.

One way to investigate this question was the use of the virtual audio 3D rendering, offering direct simulation of an individual's classical interaction with the world, allowing researchers to construct complex scenes which would be unfeasible in real contexts.

The first study reported here compared small-scale spatial configuration and immersive locomotor-scale environment on mental representation properties, and the role of auditory information in assisting the creation of spatial knowledge. Results of mental scanning times showed a clear difference in the behaviour of blind individuals following the learning of either a small or large scale environment. Although a classical positive correlation was observed between distances and mental travel times following immersion in the large environment, the small size configuration showed no time/distance correlation, either after verbal description or after tactile exploration, reflecting the poor internal organisation of their mental representation. These results suggest that, in the case of learning the spatial configuration of a locomotor space (i.e., one in which one can walk around), early deprivation of visual experience does not have a specific impact, and that early visual experience plays a dominant role in expertise regarding finer differences, supported by results obtained with the task of mental comparison of distances on small configurations. In contrast, once congenitally blind individuals learned the spatial configuration of the immersive environment, they took no more time than blindfolded sighted to compare distances. Thus, our results support the idea that the larger the initial environment, the less costly the difference in strategy between blind and sighted individuals would be, and that it is the scale of the environment, rather than the learning modality, that was critical.

The aim of the second study, using spatial audio virtual reality, was to study the ability to offer the opportunity to blind people to learn via an audio VR system, new unknown environments before a real navigation, offering the opportunity to analyse variety of acoustic cues, offsite, at their discretion, and to elaborate cognitive maps for these spaces before navigating inside.

A comparison was made between two types of learning: real on-site displacement and active navigation in a virtual architecture of two different corridor environments. For both conditions, only acoustic information was available.

Despite the low statistical power of our data due to the low number of participants, results were very informative about the question considered. Using the mental comparison of distance task, we observed the symbolic distance effect and no significant differences between performance in the real and virtual navigation conditions. This initial evaluation tended to confirm that both physical displacement (real navigation) and active virtual navigation with a joystick in a virtual architectural acoustic environment provide appropriate information for spatial understanding and allow blind individuals to create mental representations that preserve the topological and metric properties of the original environment.

These results are valuable as they offer the opportunity to blind persons to learn new environments from their own home before having to go to the real environment. Blindness leads some people to be less social, due to their fear of visiting unknown places. Our results suggest that using audio virtual reality could be of real interest for them. There was, however, a limitation due to the fact that this study was conducted along a corridor path, only allowing participants to go back and forth, linearly, and not allowing them to be free in their exploration.

That is one of the points to be addressed in the context of the RASPUTIN project. One of the research goals is to investigate if a mental map, constructed via a virtual exploration conducted off-site in the privacy of one's home, could allow blind or visually impaired persons to become well acquainted with a space, such as a new job site, a municipal office, or a museum, prior to actually setting foot in the building, thereby improving the autonomy and sense of security of the individual. Such improvements in mobility could contribute to strengthening self-confidence and increasing access to social and cultural events. To accomplish this, studies are currently being developed to examine the various aspects of the problem, from the perspectives of cognitive science (regarding spatial memory, cognitive mapping, and learning), psycho-acoustics (acoustic cues necessary for spatial comprehension), signal processing (optimisation of room acoustic audio rendering), ergonomics of technology (navigation scenarios and interactivity for improved comprehension), and, finally, with regards to improvements in the individual autonomy of visually impaired navigation (confidence, speed, and precision). Specifically with regards to cognitive maps of architectural spaces, we hope to investigate whether the mental strategies are more egocentric/navigational or allocentric/survey as a function of visual experience [76]. In sighted literature, allocentric perspective is define as global, considering a large environment, frequently termed a "bird's eye view" of the environment, and is the perspective given by maps. An egocentric perspective refers to the a representation of the environment using the body as reference; the perspective we have when we move through an environment or explore a scene [77,78]. A large body of literature [6,10] suggest that congenitally blind people would be more induced to use egocentric coding strategies when elaborating mental representation of space, and that early visual experience would allow for a more accurate understanding of external spatial coordinate systems [22,23]. The question we want to arise here, is once their mental representation of a new configuration is constructed, without inducing one or another of the strategy (egocentric vs allocentric) to resolve the task in the experimental instructions, will we be able to arrive at the same conclusions.

Finally, the transfer of information from preparatory learning to actual navigation of an architectural site will be investigated, comparing learning the layout of a building via tactile maps or virtual auditory navigation. We will investigate whether either of the cognitive maps developed in the two learning phases improves efficiency or confidence in unaided navigation in unfamiliar locations. These works are related to those of [79,80] or [81] in particular which also seek to allow a more autonomous navigation of blind people in unknown out or indoor environments, but by employing other technologies. Some works are based on the use of cell phones which have the advantage of being within reach of almost everyone [79,80] and, thus, allow the development of interactive virtual guidance applications. This work shows that participants benefit from knowledge acquired via a smartphone-based virtual navigation app on in-situ navigation tasks. However, as soon as they can benefit from a NavCog-type application on the spot, this consequently induces them to rely *more* on the system, rather than on their *a priori* knowledge, such that the use of such virtual navigation aide does not finally allow them to specifically improve their performance. Other authors have shown that navigation in virtual environments, thanks to console controllers [81] in particular, allowed blind people to retrieve information about the environment in a way that would not be possible in the real world (the look-around mode), and that this had an influence on the construction of their mental representations, which were more detailed and contained more information about the space than those of the control group. Thus, regardless of the technologies used, the goal of researchers in this field is to provide a better understanding of the processes involved in the construction of mental representations of space by blind people and how current technologies can enable them to acquire the most relevant information possible to construct reliable mental maps. All these types of works are to be encouraged for a better autonomy and social integration of disabled people.

Author Contributions: Writing—original draft preparation, A.A.-J.; writing—review and editing, B.F.G.K.; funding acquisition, B.F.G.K. All authors have read and agreed to the published version of the manuscript.

Funding: This work was funded in part by the RASPUTIN project (Room Acoustic Simulations for Perceptually Realistic Uses in Real-Time Immersive and Navigation Experiences, ANR-18-CE38-0004, rasputin.lam.jussieu.fr, accessed on 9 May 2022). Additional funding associated with the various prior projects: Wayfinding (EU STREP Grant agreement ID: 12959, cordis.europa.eu/project/id/12959, accessed on 9 May 2022), NAVIG (ANR-08-TECS-0011, anr.fr/Projet-ANR-08-TECS-0011, accessed on 9 May 2022).

Institutional Review Board Statement: Not applicable.

Informed Consent Statement: Not applicable.

Data Availability Statement: Not applicable.

Conflicts of Interest: The authors declare no conflicts of interest. The funders had no role in the design of the studies; in the collection, analyses, or interpretation of data; in the writing of the manuscript, or in the decision to publish the results

References

1. Cattaneo, Z.; Fastame, M.C.; Vecchi, T.; Cornoldi, C. Working memory, imagery and visuo-spatial mechanisms. In *Imagery and Spatial Cognition: Methods, Models and Cognitive Assessment*; Number 66 in Advances in Consciousness Research; Vecchi, T., Bottini, G., Eds.; John Benjamins Publishing Company: Amsterdam, The Netherlands, 2006; pp. 113–149. [CrossRef]
2. Schacter, D.; Nadel, L. Varieties of spatial memory: A problem for cognitive neuroscience. In *Cognitive Neuroscience*; Lister, R., Weingartner, H., Eds.; Oxford University Press: New York, NY, USA, 1991; pp. 165–185.
3. Janzen, G. Memory for object location and route direction in virtual large-scale space. *Q. J. Exp. Psychol.* **2006**, *59*, 493–508. [CrossRef]
4. Gibson, J. *The Ecological Approach to Visual Perception*; Houghton Mifflin: Boston, MA, USA, 1979.
5. Dodson-Burk, B.; Hill, E.W. *Preschool Orientation and Mobility Screening*; Division IX of the Association for Education and Rehabilitation of the Blind and Visually Impaired: Nashville, TN, USA, 1989.
6. Millar, S. Models of Sensory Deprivation: The Nature/Nurture Dichotomy and Spatial Representation in the Blind. *Int. J. Behav. Dev.* **1988**, *11*, 69–87. [CrossRef]
7. Heller, M.A.; Gentaz, E. *Psychology of Touch and Blindness*; Psychology Press: London, UK, 2013.
8. Valente, D.; Bara, F.; Afonso-Jaco, A.; Baltenneck, N.; Gentaz, É. La perception tactile des propriétés spatiales des objets chez les personnes aveugles. *Enfance* **2021**, *1*, 69–84. [CrossRef]
9. Gentaz, E. *La Main, le Cerveau et le Toucher: Approches Multisensorielles et Nouvelles Technologies*; Dunod: Ile-de-France, France, 2018.
10. Millar, S. *Understanding and Representing Space: Theory and Evidence from Studies with Blind and Sighted Children*; Clarendon Press/Oxford University Press: New York, NY, USA, 1994. [CrossRef]
11. Carpenter, P.A.; Eisenberg, P. Mental rotation and the frame of reference in blind and sighted individuals. *Percept. Psychophys.* **1978**, *23*, 117–124. [CrossRef] [PubMed]
12. Thinus-Blanc, C.; Gaunet, F. Representation of space in blind persons: Vision as a spatial sense? *Psychol. Bull.* **1997**, *121*, 20. [CrossRef]
13. Vecchi, T.; Phillips, L.H.; Cornoldi, C. Individual differences in visuo-spatial working memory. In *Imagery, Language and Visuo-Spatial Thinking*; Psychology Press: London, UK, 2000; pp. 28–58.
14. Cattaneo, Z.; Vecchi, T. *Blind Vision: The Neuroscience of Visual Impairment*; MIT Press: Cambridge, MA, USA, 2011; p. 280.
15. Afonso, A.; Blum, A.; Katz, B.F.G.; Tarroux, P.; Borst, G.; Denis, M. Structural properties of spatial representations in blind people: Scanning images constructed from haptic exploration or from locomotion in a 3-D audio virtual environment. *Mem. Cogn.* **2010**, *38*, 591–604. [CrossRef]
16. Cornoldi, C.; Cortesi, A.; Preti, D. Individual differences in the capacity limitations of visuospatial short-term memory: Research on sighted and totally congenitally blind people. *Mem. Cogn.* **1991**, *19*, 459–468. [CrossRef]
17. Gaunet, F.; Thinus-Blanc, C. Early-Blind Subjects' Spatial Abilities in the Locomotor Space: Exploratory Strategies and Reaction-to-Change Performance. *Perception* **1996**, *25*, 967–981. [CrossRef]
18. Gaunet, F.; Martinez, J.L.; Thinus-Blanc, C. Early-blind subjects' spatial representation of manipulatory space: Exploratory strategies and reaction to change. *Perception* **1997**, *26*, 345–366. [CrossRef]
19. Marmor, G.S.; Zaback, L.A. Mental rotation by the blind: Does mental rotation depend on visual imagery? *J. Exp. Psychol. Hum. Percept. Perform.* **1976**, *2*, 515–521. [CrossRef]
20. Byrne, R.W.; Salter, E. Distances and directions in the cognitive maps of the blind. *Can. J. Psychol./Revue Canadienne de Psychologie* **1983**, *37*, 293–299. [CrossRef] [PubMed]

21. Wanet, M.C.; Veraart, C. Processing of auditory information by the blind in spatial localization tasks. *Percept. Psychophys.* **1985**, *38*, 91–96. [CrossRef] [PubMed]
22. Millar, S. Self-Referent and Movement Cues in Coding Spatial Location by Blind and Sighted Children. *Perception* **1981**, *10*, 255–264. [CrossRef] [PubMed]
23. Millar, S. Movement Cues and Body Orientation in Recall of Locations by Blind and Sighted Children. *Q. J. Exp. Psychol. Sect. A* **1985**, *37*, 257–279. [CrossRef]
24. Barber, P.; Lederman, S. Encoding Direction in Manipulatory Space and the Role of Visual Experience. *J. Vis. Impair. Blind.* **1988**, *82*, 99–106. [CrossRef]
25. Carreiras, M.; Codina, B. Spatial cognition of the blind and sighted: Visual and amodal hypotheses. *Cahiers Psychologie Cognitive/Curr. Psychol. Cogn.* **1992**, *12*, 51–78.
26. Rieser, J.J.; Guth, D.A.; Hill, E.W. Sensitivity to Perspective Structure While Walking without Vision. *Perception* **1986**, *15*, 173–188. [CrossRef]
27. Loomis, J.; Klatzky, R.; Golledge, R.; Cicinelli, J.; Pellegrino, J.; Fry, P. Nonvisual Navigation by Blind and Sighted: Assessment of Path Integration Ability. *J. Exp. Psychol. Gen.* **1993**, *122*, 73–91. [CrossRef]
28. Parseihian, G.; Katz, B.F.G. Morphocons: A New Sonification Concept Based on Morphological Earcons. *J. Audio Eng. Soc.* **2012**, *60*, 409–418.
29. Chebat, D.R.; Maidenbaum, S.; Amedi, A. Navigation Using Sensory Substitution in Real and Virtual Mazes. *PLoS ONE* **2015**, *10*, e0126307. [CrossRef]
30. Chebat, D.R.; Maidenbaum, S.; Amedi, A. The transfer of non-visual spatial knowledge between real and virtual mazes via sensory substitution. In Proceedings of the International Conference on Virtual Rehabilitation (ICVR), Montreal, QC, Canada, 19–22 June 2017; pp. 1–7. [CrossRef]
31. Kosslyn, S.M.; Ball, T.M.; Reiser, B.J. Visual images preserve metric spatial information: Evidence from studies of image scanning. *J. Exp. Psychol. Hum. Percept. Perform.* **1978**, *4*, 47. [CrossRef] [PubMed]
32. Denis, M.; Zimmere, M. Analog properties of cognitive maps constructed from verbal descriptions. *Psychol. Res.* **1992**, *54*, 286–298. [CrossRef]
33. Röder, B.; Rösler, F. Visual input does not facilitate the scanning of spatial images. *J. Ment. Image* **1998**, *22*, 165–182.
34. Kerr, N.H. The role of vision in "visual imagery" experiments: Evidence from the congenitally blind. *J. Exp. Psychol. Gen.* **1983**, *112*, 265–277. [CrossRef]
35. Afonso, A.; Gaunet, F.; Denis, M. The mental comparison of distances in a verbally described spatial layout: Effects of visual deprivation. *Imagin. Cogn. Personal.* **2003**, *23*, 173–182. [CrossRef]
36. Afonso, A. Propriétés Analogiques des Représentations Mentales de l'Espace: Étude Comparative Auprès de Personnes Voyantes et Non-Voyantes. Ph.D. Thesis, Laboratoire d'Informatique pour la Mécanique et les Sciences de l'Ingénieur, Orsay, France, 2006.
37. Worchel, P. Space perception and orientation in the blind. *Psychol. Monogr. Gen. Appl.* **1951**, *65*, 1–28. [CrossRef]
38. Rieser, J.J.; Guth, M.D.A.; Hill, E.W. Mental Processes Mediating Independent Travel: Implications for Orientation and Mobility. *J. Vis. Impair. Blind.* **1982**, *76*, 213–218. [CrossRef]
39. Campos, J.J.; Anderson, D.I.; Barbu-Roth, M.A.; Hubbard, E.M.; Hertenstein, M.J.; Witherington, D. Travel broadens the mind. *Infancy* **2000**, *1*, 149–219. [CrossRef]
40. Tinti, C.; Adenzato, M.; Tamietto, M.; Cornoldi, C. Visual Experience is not Necessary for Efficient Survey Spatial Cognition: Evidence from Blindness. *Q. J. Exp. Psychol.* **2006**, *59*, 1306–1328. [CrossRef]
41. Veraart, C.; Wanet-Defalque, M.C. Representation of locomotor space by the blind. *Percept. Psychophys.* **1987**, *42*, 132–139. [CrossRef]
42. Iachini, T.; Giusberti, F. Metric properties of spatial images generated from locomotion: The effect of absolute size on mental scanning. *Eur. J. Cogn. Psychol.* **2004**, *16*, 573–596. [CrossRef]
43. Vecchi, T. Visuo-spatial Imagery in Congenitally Totally Blind People. *Memory* **1998**, *6*, 91–102. [CrossRef]
44. Katz, B.F.G.; Picinali, L., Spatial Audio Applied to Research with the Blind. In *Advances in Sound Localization*; Strumillo, P., Ed.; InTech: London, UK, 2011; pp. 225–250. [CrossRef]
45. Strelow, E.R.; Brabyn, J.A. Locomotion of the Blind Controlled by Natural Sound Cues. *Perception* **1982**, *11*, 635–640. [CrossRef]
46. Morrongiello, B.A.; Timney, B.; Humphrey, G.K.; Anderson, S.; Skory, C. Spatial knowledge in blind and sighted children. *J. Exp. Child Psychol.* **1995**, *59*, 211–233. [CrossRef] [PubMed]
47. Nicol, R. *Binaural Technology*; AES Monograph, Audio Engineering Society: New York, NY, USA, 2010; p. 77.
48. Blauert, J. *Spatial Hearing*; MIT Press: Cambridge, MA, USA, 1996.
49. Begault, D. *3-D Sound for Virtual Reality and Multimedia*; Academic Press: Cambridge, UK, 1994.
50. Katz, B.F.G.; Nicol, R. Binaural Spatial Reproduction. In *Sensory Evaluation of Sound*; Zacharov, N., Ed.; CRC Press: Boca Raton, FL, USA, 2019; pp. 349–388, ISBN 978-1-4987-5136-0.
51. Katz, B.F.G.; Dramas, F.; Parseihian, G.; Gutierrez, O.; Kammoun, S.; Brilhault, A.; Brunet, L.; Gallay, M.; Oriola, B.; Auvray, M.; et al. NAVIG: Guidance system for the visually impaired using virtual augmented reality. *J. Technol. Disabil.* **2012**, *24*, 163–178. [CrossRef]
52. Savioja, L.; Svensson, U.P. Overview of geometrical room acoustic modeling techniques. *J. Acoust. Soc. Am.* **2015**, *138*, 708–730. [CrossRef] [PubMed]

53. Svensson, P.; Kleiner, M.; Dalenback, B. Auralization—An Overview. *J. Audio Eng. Soc.* **1993**, *41*, 861–875.
54. Summers, J.E. What exactly is meant by the term "auralization?". *J. Acoust. Soc. Am.* **2008**, *124*, 697–697. [CrossRef]
55. Thery, D.; Boccara, V.; Katz, B.F.G. Auralization uses in acoustical design: A survey study. *J. Acoust. Soc. Am.* **2019**, *145*, 3446–3456. [CrossRef]
56. Thery, D.; Poirier-Quinot, D.; Jouan, S.; Katz, B.F.G.; Boccara, V. Architectural acoustic design: Observation of use cases including audio-only and multimodal auralizations. *Build. Acoust.* **2021**, *29*, 179–199. [CrossRef]
57. Postma, B.N.; Katz, B.F.G. The influence of visual distance on the room-acoustic experience of auralizations. *J. Acoust. Soc. Am.* **2017**, *142*, 3035–3046. [CrossRef] [PubMed]
58. Poirier-Quinot, D.; Katz, B.F.G. On the improvement of accommodation to non-individual HRTFs via VR active learning and inclusion of a 3D room response. *Acta Acust.* **2021**, *5*, 25. [CrossRef]
59. Postma, B.N.; Katz, B.F.G. Perceptive and objective evaluation of calibrated room acoustic simulation auralizations. *J. Acoust. Soc. Am.* **2016**, *140*, 4326–4337. [CrossRef] [PubMed]
60. Mullins, S.; Le Page, V.; De Muynke, J.; Canfield-Dafilou, E.K.; Billiet, F.; Katz, B.F.G. Preliminary report on the effect of room acoustics on choral performance in Notre-Dame and its pre-Gothic predecessor. *Meet. Acoust. Soc. Am.* **2021**, *150*, A258. [CrossRef]
61. Katz, B.F.G.; Lyzwa, J.M.; Poirier-Quinot, D. La Vierge 2020 : Reconstructing a virtual concert performance Through Historic Auralisation of Notre-Dame Cathedral. In Proceedings of the International Conference on Immersive and 3D Audio (I3DA 2021), Bologna, Italy, 8–10 September 2021; pp. 1–9. Available online: https://youtu.be/83QClpt3hyU (accessed on 9 May 2022).
62. Beni, R.d.; Cornoldi, C. Imagery limitations in totally congenitally blind subjects. *J. Exp. Psychol. Learn. Mem. Cogn.* **1988**, *14*, 650–655. [CrossRef]
63. Loomis, J.; Lippa, Y.; Golledge, R.; Klatzky, R. Spatial Updating of Locations Specified by 3-D Sound and Spatial Language. *J. Exp. Psychol. Learn. Mem. Cogn.* **2002**, *28*, 335–45. [CrossRef]
64. Afonso, A.; Katz, B.F.G.; Blum, A.; Jacquemin, C.; Denis, M. A Study of Spatial Cognition in an Immersive Virtual Audio Environment: Comparing Blind and Blindfolded Individuals. In Proceedings of the International Conference on Auditory Display, Limerick, Ireland, 6–9 July 2005; pp. 228–235.
65. Minnaar, P.; Olesen, S.K.; Christensen, F.; Møller, H. The importance of head movements for binaural room synthesis. In Proceedings of the International Conference on Auditory Display, Helsinki, Finland, 29 July–1 August 2001; pp. 21–25.
66. Frauenberger, C.; Noisternig, M. 3D Audio Interface for the Blind. In Proceedings of the International Conference on Auditory Display, Boston, MA, USA, 6–9 July 2003.
67. Jacquemin, C. *Virtual Choreographer Reference Guide (Version 1.4)*; LIMSI-CNRS and University Paris 11: Orsay, France, 2013.
68. Denis, M.; Cocude, M. Structural properties of visual images constructed from poorly or well-structured verbal descriptions. *Mem. Cogn.* **1992**, *20*, 497–506. [CrossRef]
69. Kaski, D. Revision: Is Visual Perception a Requisite for Visual Imagery? *Perception* **2002**, *31*, 717–731. [CrossRef]
70. Hegarty, M.; Montello, D.R.; Richardson, A.E.; Ishikawa, T.; Lovelace, K. Spatial abilities at different scales: Individual differences in aptitude-test performance and spatial-layout learning. *Intelligence* **2006**, *34*, 151–176. [CrossRef]
71. Ohuchi, M.; Iwaya, Y.; Suzuki, Y.; Munekata, T. A comparative study of sound localization acuity of congenital blind and sighted people. *Acoust. Sci. Technol.* **2006**, *27*, 290–293. [CrossRef]
72. Loomis, J.; Golledge, R.; Klatzky, R. Navigation system for the blind: Auditory display modes and guidance. *Presence Teleoper. Virtual Environ.* **1998**, *7*, 193–203. [CrossRef]
73. Picinali, L.; Afonso, A.; Denis, M.; Katz, B.F.G. Exploration of architectural spaces by blind people using auditory virtual reality for the construction of spatial knowledge. *Int. J. Hum. Comput. Stud.* **2014**, *72*, 393–407. [CrossRef]
74. Katz, B.F.G.; Parseihian, G. Perceptually based Head-Related Transfer Function database optimization. *J. Acoust. Soc. Am.* **2012**, *131*, EL99–EL105. [CrossRef] [PubMed]
75. Denis, M.; Cocude, M. Scanning visual images generated from verbal descriptions. *Eur. J. Cogn. Psychol.* **1989**, *1*, 293–307. [CrossRef]
76. Afonso-Jaco, A.; Zagala, F.; Katz, B.F.G. On the influence of primary visual experience and tactile map design on mental navigation across learned spatial representations. Unpublished Work, 2022.
77. Grison, E.; Gyselinck, V. La cognition spatiale pour repenser les aides à la navigation. *L'Année Psychologique* **2019**, *119*, 243–278. [CrossRef]
78. Grison, E.; Afonso-Jaco, A. Is the construction of spatial models multimodal? New evidences towards sensory-motor information involvement from temporary blindness study. *Psychol. Res.* **2021**, *85*, 2636–2653. [CrossRef]
79. Guerreiro, J.; Ahmetovic, D.; Kitani, K.M.; Asakawa, C. Virtual navigation for blind people: Building sequential representations of the real-world. In Proceedings of the 19th International ACM SIGACCESS Conference on Computers and Accessibility, Baltimore, MD, USA, 29 October–1 November 2017; pp. 280–289.
80. Guerreiro, J.; Sato, D.; Ahmetovic, D.; Ohn-Bar, E.; Kitani, K.M.; Asakawa, C. Virtual navigation for blind people: Transferring route knowledge to the real-World. *Int. J. Hum.-Comput. Stud.* **2020**, *135*, 102369. [CrossRef]
81. Lahav, O.; Gedalevitz, H.; Battersby, S.; Brown, D.; Evett, L.; Merritt, P. Virtual environment navigation with look-around mode to explore new real spaces by people who are blind. *Disabil. Rehabil.* **2018**, *40*, 1072–1084. [CrossRef]

sensors

MDPI

Article

Virtual Reality Systems as an Orientation Aid for People Who Are Blind to Acquire New Spatial Information

Orly Lahav

The Constantiner School of Education, Tel Aviv University, Tel Aviv 6997801, Israel; lahavo@tauex.tau.ac.il

Abstract: This research aims to examine the impact of virtual environments interface on the exploration process, construction of cognitive maps, and performance of orientation tasks in real spaces by users who are blind. The study compared interaction with identical spaces using different systems: BlindAid, Virtual Cane, and real space. These two virtual systems include user-interface action commands that convey unique abilities and activities to users who are blind and that operate only in these VR systems and not in real space (e.g., teleporting the user's avatar or pointing at a virtual object to receive information). This research included 15 participants who are blind, divided into three groups: a control group and two experimental groups. Varied tasks (exploration and orientation) were used in two virtual environments and in real spaces, with both qualitative and quantitative methodologies. The results show that the participants were able to explore, construct a cognitive map, and perform orientation tasks. Participants in both virtual systems used these action commands during their exploration process: all participants used the teleport action command to move their avatar to the starting point and all Virtual Cane participants explored the environment mainly by using the look-around mode, which enabled them to collect spatial information in a way that influenced their ability to construct a cognitive map based on a map model.

Keywords: blind; virtual reality; orientation and mobility; spatial perception; cognitive map

1. Introduction

People who are blind face deficits in the ability to navigate outdoor and new indoor spaces. The lack of the sense of sight makes it difficult to identify obstacles and locations independently, or simply to find a target path. Consequently, people who are blind must use compensatory multisensorial (touch, audio, and olfactory) channels and alternative exploration methods [1].

Over the past 50 years, a large number of orientation and mobility (O&M) digital technologies have been developed and researched [2]. The increased number of new O&M digital aids for people who are blind has had positive effects on O&M rehabilitation programs and their users. The new digital orientation aids provide spatial information (in advance or in situ) about unexplored spaces.

A virtual reality (VR) system can enhance the capabilities of people with sensorial, physical, mental, and learning disabilities in multiple areas [3,4]. Research and development of orientation VR for people who are blind have been conducted over the past 25 years. A survey [5] that included VR systems for people who are blind and visually impaired over the past two decades clustered the VR systems into a three-level taxonomy based on exploration interaction, perspective, application scenario, and evaluation. These VR systems compensate for lacking visual information through haptic and/or auditory feedback. The haptic feedback transmits sensation through direct interaction with the virtual object (e.g., texture and/or stiffness) to allow detection of artificial representations of real objects. The haptic devices include SensAble Phantom Desktop, Immersion Corp.'s CyberForce, and Novint Falacon, and the tactile devices include the force feedback joystick and Nintendo's Wii Controller. The auditory stimulus can include mono, stereo, or surrounding audio that allows the user to detect the direction and distance of sounds, which are then used as clues or landmarks. Past research on VR for people who are blind has revealed the benefits of such multisensorial systems. These benefits support people who are blind in perceiving spatial information, spatial problem solving, practicing and enhancing O&M skills, and building O&M strategies [6]; enabling the user's independent interaction; displaying immediate feedback suiting the user's sensory and cognitive abilities; and providing the opportunity to practice in a safe area, without time limitations or professional restrictions.

Additionally, orientation virtual environments (VEs) can support O&M specialists in the rehabilitation training process [6]. Most VR systems include indoor and outdoor spaces that allow learners who are blind to explore a new space in advance. While exploring the VE, the learner interacts with the landmarks and clues and collects spatial information that will later support him or her in constructing a cognitive map that can be applied in real space (RS). A few research teams have developed and researched orientation VEs for users who are blind, such as [7–17]. Their research findings showed that people who are blind were able to explore VE systems independently, to construct cognitive maps as a result of the exploration, and to apply this spatial knowledge successfully in familiar and unfamiliar RSs. Other orientation VR systems have been used mainly to help trainees who are blind to acquire spatial and O&M skills [9,18–21]. These research findings have indicated the potential of VR systems to play a central role in three activities: as an exploration/navigation planning tool for independent traveling in unfamiliar RSs, as a training simulator for orientation, and as a diagnostic tool for O&M specialists to track and observe learners' spatial abilities and strategies.

The VR system does have several limitations affecting all of the above uses. The VE is not a replica or replacement for a rehabilitation specialist's instruction or for exploration in RS; however, in cases in which the RS is not accessible for exploration, the VE is a good substitute.

Traditional O&M rehabilitation programs provide practice in the acquisition of O&M skills and spatial mapping, which are supplied at the perceptual and conceptual levels. At the perceptual level, people who are blind are able to collect multisensorial spatial information about their surroundings and apply it to orient themselves in indoor and outdoor spaces [22]. At the conceptual level, the focus is on developing compensatory exploration strategies to perceive a spatial representation that can be applied efficiently in RS. Spatial representation is stored as a cognitive map—a mental representation of an image based on one's knowledge about a space [23]. The cognitive map can be represented as a route model, in which the space is described in terms of a series of displacements in space; as a map model, in which the space is described from a bird's-eye view of the space; or as an integrated mental representation of both route and map models. Most O&M rehabilitation programs trainees are directed to use the route model as a spatial model to promote safety and to allow concentration on the target path.

People who are blind principally rely on a route model to construct a cognitive map [24]. They will usually require a map model upon encountering an unusable path (due to a fallen tree, road construction, etc.), engendering the need to identify an alternate route. Furthermore, people who are blind are not usually trained to explore map layers (e.g., surface, routes, traffic, public transportation, or buildings). The development of new VR systems can have value in filling these gaps by including special action commands that are unique to the VE, a benefit that is not available in RS to the independent explorer who is blind. The development of new O&M digital technologies must also take into account the challenges of developing simple and intuitive user interfaces [25], especially user interfaces accessible to people who are blind [26,27].

This study compared two orientation VR systems. Both systems contain user interfaces with unique components that enable the explorer who is blind to work independently. Both convey to the user abilities and activities that operate only in these VR systems and are not available in RS. The two VR systems, the BlindAid system [9,28] and the Virtual Cane (Wiimote) [29], have been previously studied. In this research, we examine and compare the spatial behavior of participants who are blind exploring two multisensorial VR systems. The VEs were identical and represented corresponding simple and complex RSs. Each group of participants explored one VR system or RS. The two main goals of the research were: first, to understand the impact of a multisensorial VE on spatial abilities for users who are blind by comparing their exploration in identical unfamiliar spaces using different VR systems or RS; and second, to examine the use of unique user-interface action commands and their influence on the exploration process, the construction of cognitive maps, and the transfer of this spatial knowledge during O&M in the RS. This study explores three research questions:

1. How do people who are blind explore unfamiliar spaces using BlindAid or Virtual Cane, or in RS?
2. What were the participants' cognitive mapping characteristics after exploration using the BlindAid or Virtual Cane, or in RS?
3. How did the control group and the two experimental groups (BlindAid and Virtual Cane) perform the RS orientation tasks?

2. Materials and Methods

2.1. Participants

This research included 15 participants; six criteria were used to choose the research participants: totally blind without residual vision; having onset of blindness at least two years prior to the research period; without other disabling condition; instructed previously in O&M; understanding English; and familiarity with computer use. We defined three groups: two experimental groups, BlindAid and Virtual Cane (Wiimote), and control. All research groups were comparable in age, gender, age of vision loss, and ability to use mobility devices. Each research group was composed of five participants (Table 1). The BlindAid experimental group participants explored the unfamiliar spaces visiting VEs through a Phantom device. The Virtual Cane experimental group participants explored the unfamiliar spaces as VEs with a Wii controller. In contrast, the control group participants investigated the unfamiliar spaces by exploring the RSs. All participants independently explored the unfamiliar RSs. To recruit the participants, we used snowball sampling; each participant was randomly assigned to a research group. Each participant completed an O&M questionnaire to assess O&M abilities. The O&M questionnaire outcomes revealed no differences in O&M ability in any group: in familiar indoor spaces (home or work), none of the participants used a mobility aid; in familiar indoor spaces (small or large shopping mall), 61% of the participants chose to be escorted by a sighted person; in familiar outdoor spaces (their neighborhood with street crossing and public transportation) 60% of the participants used a mobility device (white cane or guide dog); in familiar crowded outdoor and unfamiliar indoor spaces all the participants chose to use a mobility device or to be escorted by a sighted person; in unfamiliar indoor spaces, such as shopping areas, and in unfamiliar outdoor spaces, 88% of the participants chose to be escorted by a sighted person.

Table 1. Research participants.

	Age Mean	Gender		Age Of Vision Loss		O&M Aids	
		Female	**Male**	**Congenitally**	**Adventitiously**	**White Cane**	**Guide Dog**
BlindAid experimental group (n = 5)	43 (28–59)	2	3	5	0	4	1
Virtual Cane experimental group (n = 5)	30 (25–40)	3	2	3	2	3	2
Control group (n = 5)	40 (27–56)	1	4	2	3	3	2

2.2. Variables

The independent variable in this research was the degree of complexity of spaces explored by the participants (including a simple and a complex space). This level of complexity was related to the size of the space, its structure, and the number of components within it. An O&M rehabilitation specialist examined these spaces to address safety and O&M issues. Three groups of dependent variables were defined: the exploration process, construction of a cognitive map, and performance of orientation tasks in RS. These variables have been defined in our previous research [8,28]. The exploration process included five variables: (1) duration; (2) modes: walk-around mode (exploring the space by walking) or look-around mode (standing in the space in one spot and gathering information as requested about the names of the structure's components, distance between the structure's components, objects' name, and objects' distance); (3) spatial strategies: random, exploring object area, object-to-object, grid, and perimeter; (4) length of pauses not resulting from a technical issue; and (5) the use of orientation aids in the VE (e.g., teleport action command) or the RS (e.g., using the second hand). Four variables were studied in the building of a cognitive map: (1) components: objects, objects' location, structural component, and structural components' location; (2) spatial strategy used for describing the space: starting-point perspective descriptions, items list, object-to-object, or perimeter; (3) spatial representation model used for describing the space: route model, map model, and integrated representation of route and map models; and (4) chronology of the descriptive process. Three variables were related to the orientation tasks performance: (1) the response time of correct orientation task performance (RTC); (2) successful completion of finding the task's target: arrival at the target (or at the target's zone), arrival at the target's zone with verbal assistance, or failure to arrive; and (3) type of path: direct, direct with limited walking around, indirect, or wandering around.

2.3. Instrumentation

Three implementation tools and five data collection tools were used. The three implementation tools were:

Real spaces. Two RSs situated on a university campus were used. Two unfamiliar indoor spaces were selected to show how exploration in the VE or RS affected the acquisition of spatial knowledge by people who are blind. The simple space (Figure 1) was a rectangle of 44 square meters containing two windows, five doors, and nine objects (dark green): a communications cabinet (1), an electric cabinet (2), two mailboxes (3–4), a chair (5), a bench (6), a recycling bin (7), and two boxes (8–9). The complex space (Figure 2) was larger and contained two long parallel hallways with two short parallel hallways connecting the long ones; and 12 objects: two benches (1, 7), staircases (2, 16), recycling bin (3), snack machine (4), two round tables (5–6), chair (8), mailbox (9), window (12), electric cabinet (13), pole (14), and box (15).

The BlindAid. Development and research of the BlindAid system took place at the MIT Touch Lab as part of a collaborative research project [6,28]. The BlindAid system permits people who are blind to manipulate virtual objects and provides multisensorial feedback. The VE software runs on a personal computer equipped with a haptic device—a Desktop Phantom device (SensAble Technologies, Woburn, MA, USA) and stereo headphones (Sennheiser HD580, Wedemark, Germany) (Figure 3). The Desktop Phantom device provides haptic feedback from the tip of the Phantom, similar to that generated by the tip of a white cane (e.g., stiffness and texture). The haptic device can simulate different degrees of ground texture and stiffness (e.g., marble floor or rubber floor) and the textures and stiffness of different objects (e.g., table or sofa). The system includes surrounding audio feedback that conveys sounds to the users as if they were standing in the VE.

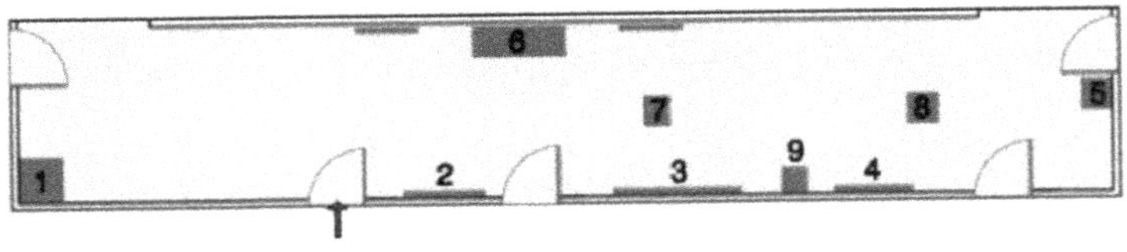

Figure 1. Simple space.

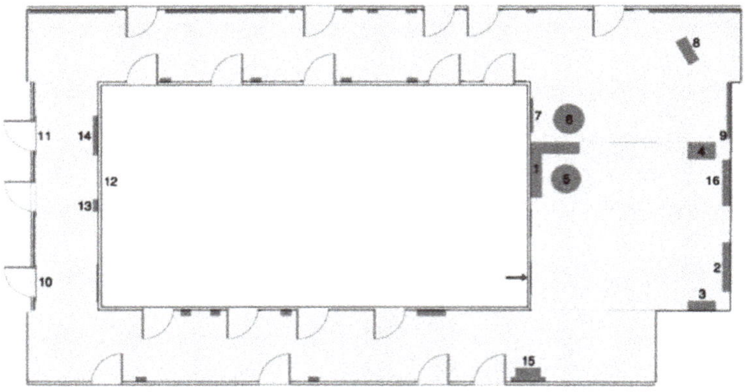

Figure 2. Complex space.

Figure 3. BlindAid system.

The virtual workspace is a rectangular box that corresponds to the usable physical workspace of the Phantom, and the user avatar is always contained within the workspace. To move the virtual workspace within the VE in order to explore beyond the confines of the workspace, the user presses one of the arrow keys. Each arrow key press shifts the workspace half of its width in the given direction. The VR system includes unique action commands that are available to the user only in this VR system and not in the RS. The six action commands on the computer's numeric keypad include teleport, pause, start, additional audio information, exploring the VE's structure layer without the objects in it, and exploring the VE's structure with the objects. Further technical details about the system were presented in our earlier paper [30]. For this study, eight VEs were designed to train the participants on the use of the BlindAid system (Figure 4). These eight VEs differed in their level of complexity: size, shape, components (structure and objects), and components' location.

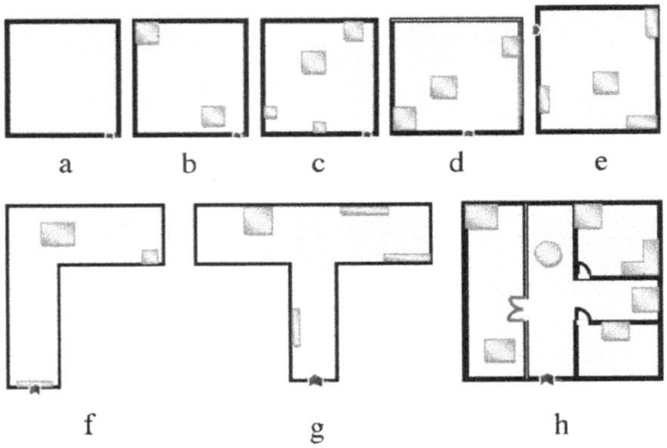

Figure 4. VE training spaces: (**a**) an empty squre room; (**b**) a squre room with one object attached to wall, and second located far from the wall; (**c**) a squre room with three objects in different sizes attached to the wall, and one more located far from the wall; (**d**) a rectangle room with two objects in different sizes attached to the wall, and one more located in rooms' center; (**e**) a rectangle room with three objects attached to the wall, and one more located in rooms' center; (**f**) a "L" shape space with three objects: a door, one object attached to the wall, and one more located in spaces' center; (**g**) a "T" shape space with five objects: a door and four objects in different sizes attached to the wall; and (**h**) a squre space devided to three rooms with a corridoe in "T" shape with a door and seven objects in different sizes and locations.

To evaluate the participants' spatial ability, two additional VEs were designed in the BlindAid system based on the RSs that were chosen earlier: a simple space (Figure 1) and a complex space (Figure 2). These simple and complex VEs are identical to the RSs and to the Virtual Cane VEs in layout and components.

The Virtual Cane. The Nintendo Wii controller (Wiimote) system was developed in collaborative research with the team at the Computing and Informatics Research Centre at Nottingham University. The Wii technology is popular, low cost, and easy to use with a standard personal computer (PC) [31] (Figure 5).

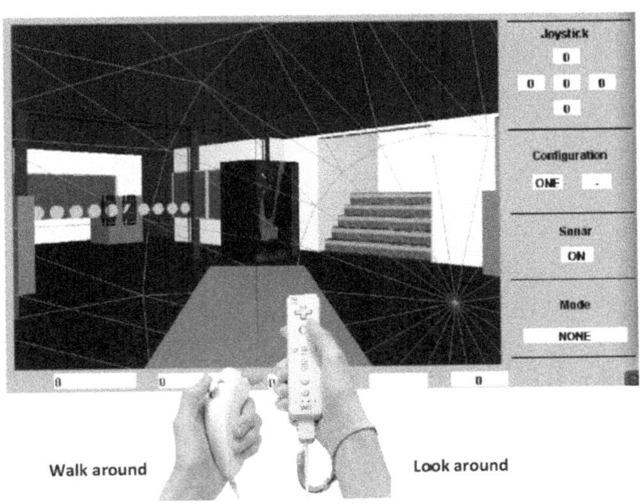

Figure 5. Virtual Cane system.

The Wiimote includes tracking technologies (e.g., accelerometer and infrared camera); the Wiimote system permits people who are blind to interact with the remote controller (Figure 6) as a white cane and in addition as a handheld camera in the VE. The VE interface was developed by the Windows-based Wii Controller Interface Suite (WiiCi). The WiiCi tools enable a connection between a personal computer and controller [32].

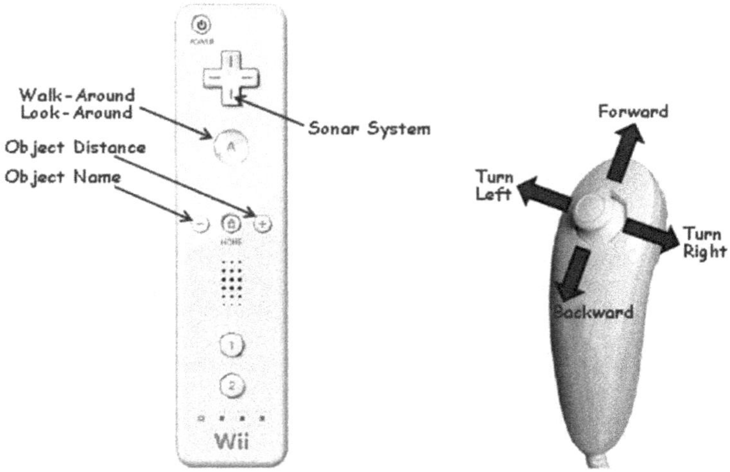

Figure 6. The Wiimote remote controller (**left**) and Nunchuck controller (**right**).

The Wiimote system was connected to a PC and the participants were seated next to it. The participant held the Wiimote remote controller and the Nunchuck (Figure 6) in both hands and wore stereo headphones (Sennheiser HD580). The Wiimote remote controller has five buttons: button "+" produces audio feedback about the object distance; button "–" produces audio feedback about the object name; button "A" toggles from walk-around to look-around mode; the arrows button operates the sonar system; and button "Home" offers an auditory description of the space.

The Virtual Cane system operates in two modes:

Look-around mode. The look-around mode is realized through the movement and orientation of the Wiimote remote controller. The avatar point of view is directly slaved to that of the remote controller. Users receive auditory and tactile (vibration) feedback by using the remote controller for scanning for objects in front of them. Each structure or object component in the VE has unique auditory and tactile feedback. The audio feedback includes the object's name and its distance. Tactile feedback vibration is activated and differs in accordance with the distance to an object (determined via ray-casting from the point of reference of the virtual Wii controller). A constant rumble is triggered by a collision with an object or the structure's components. When the user is out of the look-around mode, tactile feedback is given as a constant rumble. Moving the avatar from one area to another produces a whooshing sound. The look-around mode is available only in this orientation VR system, which was designed especially for users who are blind.

The walk-around mode. The walk-around mode is operated by tilting the Nunchuck controller in four directions (left, right, forward, or back). With the tilt direction of the Nunchuck controller, the user avatar moves in a fixed 15-degree turn in the VE. The walking speed is related to the severity of tilt. Auditory feedback (right or left) is heard before the activation of a turn. Furthermore, auditory feedback indicating footsteps is received as the avatar walks in the VE. Further technical details about the Virtual Cane system can be found in our earlier paper [28].

To allow users to train themselves in operating this Virtual Cane system, eight VEs were built (identical to the BlindAid training VEs, Figure 4). Further, simple and complex

VEs were designed identical to the RSs (Figures 1 and 2) and BlindAid VEs in layout and components.

The five data collection tools are as follows:

O&M questionnaire. The questionnaire is a self-evaluation of O&M skills and abilities. This questionnaire included 50 questions regarding the participant's O&M abilities in familiar and unfamiliar indoor and outdoor spaces. The O&M questionnaire was the same as employed in previous research [6,8,28,29]. Four O&M rehabilitation specialists assessed the questionnaire.

Exploration task. All participants were invited to explore both spaces (simple and complex spaces) in a limited timeframe that was recommended by an O&M rehabilitation specialist (40 min for exploring the simple space and 60 min for exploring the complex space). The control group explored the RSs and both experimental groups explored them via the virtual systems (BlindAid or Virtual Cane).

Verbal description task. After the exploration task, the participants were asked to give a verbal description of the space. The verbal description served as an instrument to measure the cognitive map that the participants constructed as a result of their exploration in the RS or VE. The description tasks were video recorded and transcribed.

RS orientation tasks. Following the exploration and verbal description tasks, the participants were asked to perform orientation tasks in the RSs (simple and complex): (1) object-oriented tasks—the participants were requested to travel from the original exploration starting point (exploration task) to an object (three tasks in the simple space and two tasks in the complex space); (2) perspective-change tasks—the participants were asked to go from a new starting point to an object (three tasks in the simple space and two tasks in the complex space); and (3) point-to-the-location tasks—the participants were located in the original exploration starting point and were requested to point with their hand at the location of six different structural or object components.

Observations. Screen recordings were made of the experimental groups' activities in the BlindAid or Virtual Cane (Wiimote) systems and videos of the participants were recorded. Synchronization of both video recordings was performed on the researcher's computer using Camtasia 2 (screen recording software). The control group participants were video recorded in the RS exploration tasks. All research groups were video recorded during their verbal description and RS orientation tasks.

2.4. Data Analysis

The data were analyzed employing quantitative and qualitative methodologies. To assess the participants' O&M performance in the exploration task, verbal description task, and RS orientation tasks, we used coding schemes that were principally developed in previous research and evaluated by four O&M rehabilitation specialists [6,8,9,29]. We developed coding schemes for each task based on the previous coding schemes analyses and the O&M literature [33,34]. Data analysis was performed using Microsoft Excel® and Mangold Interact® software.

2.5. Procedure

The research included three to seven sessions: in the first session, all participants signed the consent form and answered the O&M questionnaire. In the second session, all participants individually explored the VEs or RSs. The experimental group participants had a total of six sessions: four sessions were devoted to learning to operate the VE system, and two sessions were spent exploring the simple and complex VEs; the duration of each session was 90 min. The control group participants had two sessions to explore the simple and complex spaces; as for the experimental groups, the duration of each session was 90 min. Following each exploration task (simple and complex space), all participants verbally described the space and then performed the RS orientation tasks. All participants performed all tasks in the same protocol order.

2.6. Research Limitation

The limitation of this study to 15 subjects arose from its exploratory nature and the challenges of participant recruitment. The small sample prevented the running of statistically significant tests, but the data collected reveal interesting distinctions that can be further evaluated with a larger sample in the future.

3. Results

Research Question 1: How do people who are blind explore unfamiliar spaces using BlindAid or Virtual Cane, or in RS?

Both experimental groups (BlindAid and Virtual Cane (Wiimote) systems) and control group (RS) participants explored the VE or the RS independently. In the simple space, all the participants performed the exploration task in less than the suggested exploration time (40 min). The BlindAid experimental group took an average of 00:19:43 min. The Virtual Cane experimental group took an average of 00:41:44 min. Three participants performed in less than the suggested exploration time (40 min), and the other two participants needed more time. The control group took an average of 00:04:39 min, about seven times less than the suggested exploration time (30 min). The participants in the BlindAid and the control groups were able to explore the spaces only by the walk-around mode (by choosing one of the spatial strategies such as perimeter, object-to-object, and other). In contrast, the Virtual Cane experimental group participants were able to implement both look-around and walk-around modes (Table 2). The BlindAid participants used the walk-around mode in 98% of the exploration time, mainly the perimeter strategy (95%), and used 2% (00:00:24) of their exploration time for pauses. The results showed that the Virtual Cane participants mainly used the look-around mode to explore the VE (74%), walked around in the VE for only 9% of the time, and paused for 17% of the time (00:07:06). Their most used walk-around mode spatial strategy was the object-to-object strategy. The control group participants who explored the RS used the walk-around mode for 94% of their exploration time, by using mostly the perimeter strategy (84%); only 6% (00:00:17) of their exploration time was used for pauses.

Table 2. Exploration process in simple space.

	n	Duration (min)	Look-Around Mode		Walk-Around Mode Spatial Strategy			Pauses (min; %)	Second Hand
			Name	Distance	Perimeter	Object-to-Object	Other		
BlindAid experimental group	1	31:00	NA	NA	100%	0%	0%	0%	NA
	2	24:30	NA	NA	99%	0%	0%	0%	NA
	3	4:56	NA	NA	98%	0%	1%	0%	NA
	4	28:02	NA	NA	90%	0%	0%	02:31; 9%	NA
	5	10:08	NA	NA	86%	0%	12%	00:06; 1%	NA
	Mean	19:43	NA		95%	0%	3%	00:24; 2%	NA
Virtual Cane experimental group	1	29:24	37%	33%	2%	6%	0%	06:46; 23%	NA
	2	49:39	28%	54%	5%	6%	0%	03:29; 7%	NA
	3	38:32	41%	33%	0%	3%	0%	08:29; 22%	NA
	4	58:45	45%	21%	4%	3%	0%	16:27; 28%	NA
	5	32:18	50%	28%	6%	8%	0%	02:16; 7%	NA
	Mean	41:44	74%		3%	5%	0%	07:06; 17%	NA
Control group	1	3:44	NA	NA	87%	0%	3%	00:22; 10%	90%
	2	11:27	NA	NA	62%	2%	29%	00:48; 7%	97%
	3	2:01	NA	NA	81%	0%	6%	00:16; 13%	72%
	4	1:26	NA	NA	100%	0%	0%	0%	7%
	5	4:38	NA	NA	89%	11%	0%	0%	63%
	Mean	4:39	NA		84%	3%	8%	00:17; 6%	52%

To explore the complex VE the BlindAid participants took an average of 00:36:46 min; the Virtual Cane experimental group took an average of 00:53:19 min; and the control group participants took an average of 00:09:14, about four to six times less compared with the BlindAid and Virtual Cane groups (Table 3). To explore the complex space, the BlindAid experimental group mainly used the perimeter strategy (96%), with very few pauses during the exploration (00:00:44; 2%). The Virtual Cane experimental group performed the tasks by using mostly the look-around mode (73%), using less the walk-around mode (12%), and making use of long pauses (15%; 00:08:00). The most-used spatial strategies were the object-to-object strategy (11%) and perimeter (9%). The control group participants explored the RS by walking for 99% of their exploration time, mainly using the perimeter strategy (82%), with pauses taking only 1% (00:00:05) of their exploration time.

Table 3. Exploration process in complex space.

	n	Duration (min)	Look-Around Mode		Walk-Around Mode Spatial Strategy			Pauses (min; %)	Second Hand
			Name	Distance	Perimeter	Object-to-Object	Other		
BlindAid experimental group	1	22:45	NA	NA	90%	1%	0%	01:36; 7%	NA
	2	73:04	NA	NA	99%	0%	0%	0%	NA
	3	13:16	NA	NA	100%	0%	0%	0%	NA
	4	55:34	NA	NA	92%	0%	3%	01:07; 2%	NA
	5	19:10	NA	NA	100%	0%	0%	0%	NA
	Mean	36:46	NA		96%	0%	1%	00:44; 2%	NA
Virtual Cane experimental group	1	26:36	51%	22%	1%	7%	0%	05:35; 21%	NA
	2	63:11	26%	49%	3%	13%	0%	03:47; 6%	NA
	3	44:43	46%	32%	2%	2%	0%	07:09; 16%	NA
	4	86:53	43%	23%	5%	6%	0%	19:07; 22%	NA
	5	45:13	51%	20%	9%	11%	0%	03:10; 7%	NA
	Mean	53:19	73%		4%	8%	0%	08:00; 15%	NA
Control group	1	7:18	NA	NA	78%	6%	15%	0%	76%
	2	19:23	NA	NA	82%	0%	18%	0%	100%
	3	10:05	NA	NA	84%	3%	9%	0%	4%
	4	3:01	NA	NA	100%	0%	0%	0%	4%
	5	6:24	NA	NA	68%	0%	23%	00:35; 9%	60%
	Mean	9:14	NA		82%	2%	13%	00:05; 1%	43%

A comparison of spatial behavior in the simple and complex spaces shows that there was an almost equal division of exploration time among the three groups. This comparison highlights four main differences: exploration duration, spatial strategies, pauses, and exploration aids. The BlindAid experimental group took four times longer than the control group to explore each space; similar results were found with the second experimental group—using the Virtual Cane took six to nine times longer compared with the control group. The choice of exploration mode and spatial strategies depended upon which space was being explored (BlindAid, Virtual Cane, or RS). The participants in the BlindAid experimental group and control group managed to use only the walk-around mode and they mainly used the perimeter strategy, while the Virtual Cane experimental group mostly chose to use the look-around mode and object-to-object strategy in the walk-around mode, with the perimeter as their secondary strategy. During the exploration of both spaces, the BlindAid (00:00:24; 00:00:44) and control group (00:00:17; 00:00:05) participants tended to use shorter pauses compared to the Virtual Cane participants (00:07:06; 00:08:00).

During the exploration process, a variety of exploration aids (action commands) was used, depending on the VE or the RS. The BlindAid system allows its participants to use three action commands: teleport (move the user's avatar to the starting point), obtain additional auditory information, and explore the VE with or without objects in it. The Virtual Cane system allows the use of the teleport action command and the look-around and walk-around modes to obtain name or distance information. In the RS the participants walked to the starting point and used their second hand to explore the space.

The results show that 66% of the BlindAid participants used the teleport action command, all the participants used the additional auditory information action command, and only one participant chose to explore the complex VE without objects for the first half of her exploration duration. All the Virtual Cane experimental participants used the teleport action command to move the user's avatar to the starting point during their exploration. Using look-around mode, the participants asked for four different types of auditory information: object name, object distance, structure component name, and structure component distance. The results showed that, while using the look-around mode, for 40% to 43% of the time (simple vs. complex spaces), the participants requested auditory feedback about the object or structure component's name, and 34% to 29% (simple vs. complex) of the time they asked for information about the object or structure component's distance (by pointing with the Wiimote remote controller at virtual components in the VE, an audio feedback was received describing the number of steps between the avatar and the pointed virtual components). Additionally, in a scenario in which the participants collided with a virtual component, they selected distance information rather than its name. Although the Virtual Cane allows the use of the look-around mode to indicate heights, only one participant in the simple space made use of this action command and then only for a few seconds.

The control group participants used their second hand to explore the RS during their exploration time in the simple (52%) or complex (43%) spaces. Only two participants chose to return to the starting point.

Research Question 2: What were the participants' cognitive mapping characteristics after exploration using the BlindAid or Virtual Cane, or in RS?

After each exploration task, all the participants were asked to verbally describe the space. We evaluated their verbal descriptions with four variables: space components (object, object location, structural component, and structural component location); spatial strategy; spatial representation model; and chronology description.

We examined verbal descriptions of the simple (Table 4) and the complex (Table 5) spaces. In their description of the simple space, the BlindAid experimental participants described an average of 59% of the total components (structure and objects) that were placed in the simple space; the Virtual Cane participants described an average of 69% of the total components, compared to an average of 40% by the control participants. All the research participants included more objects than structural components in their verbal descriptions. The participants used all types of spatial strategies to describe the space (e.g., perimeter, object-to-object, starting point, area, and list). The main difference was found in the spatial representation that was used during the verbal description: the BlindAid experimental group and control group participants mainly used a route model, compared to the Virtual Cane experimental group participants, who employed a map model. Most descriptions began with mention of a structural component, except for one participant from the BlindAid experimental group, who started his description with the content description.

Table 4. Verbal description process of simple space.

	n	Space Components	Spatial Strategy	Spatial Representation	Chronology
BlindAid experimental group	1	65%	Perimeter	Route model	Structure
	2	87%	Perimeter	Route model	Structure
	3	26%	List	List	Content
	4	78%	Perimeter	Route model	Structure
	5	37%	Area	Route model	Structure
	Mean	59%		4 Route model; 1 List	4 Structure; 1 Content
Virtual Cane experimental group	1	66%	List	Map model	Structure
	2	79%	Starting point	Map model	Structure
	3	68%	Perimeter	Route model	Structure
	4	72%	Perimeter & object to object	Map model	Structure
	5	62%	Starting point & object to object	Map model	Structure
	Mean	69%		4 Map model; 1 Route model	5 Structure
Control group	1	36%	Perimeter	Route model	Structure
	2	54%	Area	Map model	Structure
	3	27%	Starting point	Route model	Structure
	4	15%	List	List	Structure
	5	70%	Starting point	Route model	Structure
	Mean	40%		3 Route model; 1 Map model; 1 List	5 Structure

Table 5. Verbal description process of complex space.

	n	Space Components	Spatial Strategy	Spatial Representation	Chronology
BlindAid experimental group	1	42%	Perimeter & list	Route model	Structure
	2	63%	Perimeter	Route model	Structure
	3	50%	Perimeter	Route model	Structure
	4	53%	Perimeter & area	Route model	Content
	5	28%	Area	List	Structure
	Mean	47%		4 Route model; 1 List	4 Structure; 1 Content
Virtual Cane experimental group	1	20%	List	List	Structure
	2	46%	Area	Route model	Structure
	3	41%	Perimeter	Route model	Structure
	4	53%	Perimeter & object to object	Map model	Structure
	5	59%	Area	Map model	Structure
	Mean	44%		2 Map model; 2 Route model; 1 List	5 Structure
Control group	1	30%	Starting point	Map model	Structure
	2	50%	Area	Map model	Structure
	3	30%	Object to object	Route model	Structure
	4	14%	Perimeter & starting point	Route model	Content
	5	27%	Perimeter & starting point	Route model	Structure
	Mean	30%		3 Route model; 2 Map model	4 Structure; 1 Content

For the complex space (Table 5), the BlindAid experimental participants mentioned in their description an average of 47% of the total components that were located in the VE and the Virtual Cane participants mentioned an average of 44% of the total components, compared to 30% for the control participants. The participants included more objects than structural components in their verbal descriptions and used all types of spatial strategies to describe the space. In the BlindAid experimental group's verbal descriptions, four participants employed a route model and one participant listed them. Verbal descriptions by the Virtual Cane experimental group included the use of a map model by two participants, use of a route model by two participants, and list by one participant; similar results were found in the control group, where a map model was used by two participants and a route model by three participants. As they did in the simple space, here, most of the participants began their verbal descriptions with a structural component, except for two participants from the BlindAid experimental group and the control group (each), who first described the objects.

The results show differences among the three research groups and in both spaces. Participants in both experimental groups (in both spaces) included greater detail in their verbal descriptions compared to the control group participants. The BlindAid experimental group and control group described both spaces using mainly a route model, as opposed to two participants from the control group who used the map model in the complex space. In contrast, the Virtual Cane experimental group mainly described the simple space using a map model, while in describing the complex space, two participants constructed a map model.

Compared with the complex space, the simple space was described in more detail by all participants.

Research Question 3: How did the control group and the two experimental groups (BlindAid and Virtual Cane) perform the RS orientation tasks?

To answer the third question, the performance of orientation tasks by participants in the RSs was evaluated. Orientation tasks in the simple space included three object-oriented tasks, three perspective-change tasks, and one point-to-the-location task (Table 6), and in the complex space, two object-oriented tasks, two perspective-change tasks, and a point-to-the-location task (Table 7). These tasks were evaluated by the response time of correct (RTC) orientation task performance, success, and the path that was chosen.

Table 6. Success in orientation tasks in simple space.

	n	Object-Oriented			Perspective-Change			Point-to-the-Location
		RTC (s)	Success	Direct Path	RTC (s)	Success	Direct Path	
BlindAid experimental group	1	9	67%	67%	37	100%	100%	83%
	2	18	100%	100%	53	67%	67%	33%
	3	26	100%	100%	25	100%	100%	100%
	4	194	33%	0%	92	67%	67%	50%
	5	NA	67%	33%	NA	100%	100%	67%
	Mean	62	73%	60%	52	87%	87%	67%
Virtual Cane experimental group	1	57	100%	100%	192	100%	67%	83%
	2	24	67%	67%	129	33%	0%	67%
	3	34	67%	67%	68	67%	67%	83%
	4	39	67%	67%	74	67%	33%	67%
	5	68	67%	33%	93	67%	67%	100%
	Mean	44	74%	67%	111	67%	47%	80%
Control group	1	7	33%	33%	27	33%	33%	67%
	2	10	100%	100%	27	100%	100%	100%
	3	17	100%	100%	9	100%	100%	100%
	4	9	100%	100%	21	67%	67%	83%
	5	10	33%	33%	22	67%	67%	17%
	Mean	11	73%	73%	21	73%	73%	73%

Table 7. Success in orientation tasks in complex space.

	n	Object-Oriented			Perspective-Change			Point-to-the-Location
		RTC (s)	Success	Direct Path	RTC (s)	Success	Direct Path	
BlindAid experimental group	1	NA	0%	0%	NA	50%	50%	67%
	2	NA	100%	100%	NA	0%	0%	33%
	3	122	50%	50%	25	100%	100%	100%
	4	43	50%	50%	0	0%	0%	67%
	5	237	100%	100%	185	100%	50%	83%
	Mean	134	60%	60%	105	50%	40%	70%
Virtual Cane experimental group	1	0	0%	0%	333	100%	100%	33%
	2	0	0%	0%	132	50%	0%	17%
	3	194	50%	0%	117	50%	50%	50%
	4	185	100%	50%	320	100%	50%	0%
	5	185	50%	0%	0	0%	0%	50%
	Mean	188	40%	10%	180	60%	40%	38%
Control group	1	67	50%	0%	72	50%	50%	33%
	2	44	100%	100%	69	100%	50%	100%
	3	19	50%	50%	42	100%	100%	67%
	4	175	100%	50%	60	100%	100%	50%
	5	0	0%	0%	95	50%	0%	17%
	Mean	76	60%	40%	68	80%	60%	53%

In the simple space, in both experimental groups, participants took four to six times more time to successfully perform the object-oriented tasks (Table 6), and two to five times more time to perform the perspective-change tasks successfully. Success in arriving at the target for object-oriented tasks was similar among the three research groups (73–74%). In perspective-change tasks, 87% of the BlindAid experimental participants, 67% of the Virtual Cane experimental participants, and 73% of the control participants were successful. Regarding the type of path, in the object-oriented tasks, 60–67% of both experimental groups walked directly to the target, compared to 73% for the control group; in perspective-change tasks, 87% of the BlindAid experimental participants, 47% of the Virtual Cane experimental participants, and 73% of the control participants walked directly to the target. In the point-to-the-location task, most of the participants in the three groups were able to point accurately to the target objects (67% of the BlindAid participants, 80% of the Virtual Cane participants, and 73% of the control participants).

In the complex space (Table 7), the experimental group's participants took two times more time to complete successfully the object-oriented tasks, and two to three times more time to complete the perspective-change tasks successfully. Two of the Virtual Cane experimental participants stayed in the entrance lobby and did not explore the entire VE.

These limitations in exploration affected performance in the orientation RS tasks. In their first object-oriented task, they tried to transfer their VE landmarks to the real complex space to ground their spatial knowledge. Following the first object-oriented task, they became more self-assured, as also seen in the task durations, which became shorter even when a target object was farther away, and also in the perspective-change tasks. In performing object-oriented tasks, similar results were found among the BlindAid and control groups in their success in arriving at the target (60%), with 40% success for the Virtual Cane experimental group. Their performance in perspective-change tasks varied, with 80% success by the control group, 60% by the Virtual Cane experimental participants, and 50% of the BlindAid experimental participants in arriving at the target. Regarding the type of path in the object-oriented tasks, 60% of the BlindAid experimental group walked directly to the target, compared to 10% for the Virtual Cane experimental group, and 40% for the control group. In perspective-change tasks, 60% of the control group and 40% of both experimental groups walked directly to the target. In the point-to-the-location task, the BlindAid experimental participants pointed successfully at a rate of 70%, while 38% of the Virtual Cane experimental group succeeded in this task.

In the point-to-the-location task, differences were found among the three research groups. Most of the BlindAid experimental participants (70%) were able to point accurately to the target objects, while 38% of the Virtual Cane participants were able to point accurately; one participant failed as a result of a mirror distortion, and there was only 17% success in the control group.

This research took place in a real campus, which includes sounds that are particular to a crowded university setting. A selection of these sounds occurred in the VE and served as landmarks for the participants, for example, the sound of a snack machine or elevator. The experimental participants used these auditory landmarks eight times compared to only four times by the control participants.

4. Discussion

This study followed previous research that examined the use of a multisensorial VR to perceive spatial information in a VE using the Virtual Cane system [28]. The previous research results showed that exploring VEs through the look-around model influenced the spatial ability of the participants to construct a cognitive map based on the map model. To determine the influence of the look-around mode on the participants' spatial ability, this study compared the spatial ability of two experimental research groups using different VR systems exploring the same spaces. The first experimental group used the BlindAid system, which only allows users to walk in the VE to explore the space (similar to RS), and the Virtual Cane system, which offers users the choice of look-around or walk-around modes to explore the space.

This discussion addresses the research goals, which focus on the impact of exploring multisensorial VE systems or RS on the spatial abilities of people who are blind, and the use of unique user-interface action commands and their influence on the exploration process, the construction of cognitive maps, and the application of this spatial knowledge in orientation in the RS. To address these aims, we designed two Ves, one simple and one complex, which were exact representations of the RSs, for both VR systems. Both spaces were unfamiliar to all research participants.

4.1. The Impact of Multisensorial VE Systems on the Spatial Abilities of People Who Are Blind

This research paradigm involved three phases: exploring unfamiliar space, constructing a cognitive map, and performing orientation tasks in the RS [35]. These three phases involved the transformation of spatial abilities and spatial information from the RS to the VE and vice versa. The exploration in the multisensorial Ves and the RS was based on previously acquired spatial skills and strategies, which focused on how to explore and collect spatial landmarks and clues in unfamiliar spaces. Later, these explorations assisted participants in constructing a cognitive map. This spatial information (from the exploration process and cognitive map) was transferred from the VE to the RS during RS orientation tasks. Results drawn from the constructed cognitive maps show that all participants who explored the VEs included more details of the components (structure and objects and their location) compared with the participants who explored only the RSs. In regard to the RS orientation tasks performance, three aspects were of interest: RTC, successful completion of finding the task's target and point-to-the-location, and type of path. The participants in the experimental groups required more time to perform all the tasks. In regard to their success in finding the task's target and point-to-the-location, participants who explored the VEs successfully performed most of the RS orientation tasks (50%) or were equally successful (33%) compared with the control group (17%). In choosing the direct path to their target, the results between the experimental groups and control group were equal (50%). These performance-success results demonstrate that exploration through multisensorial VR systems results in spatial ability at a level better or equal to that achieved in RS exploration. In addition, VE explorers who are blind will need a longer time to find their path in the RS, as this will be their first time walking in the RS after exploring the VE only. Over time, with more practice and frequency of use of VE exploration, this length of time might drop. These results highlight the need for such an orientation tool, especially when independent exploration of unfamiliar space is not possible; in this event, multisensorial VE systems can substitute for the RS.

Similar results were found previously in orientation VR system research [3,6,7,10,11,14–17,19–21,28]; all research participants were able to explore unfamiliar space independently.

4.2. The Impact of Unique Action Commands Embedded in Multisensorial VR Interface on the Spatial Abilities of People Who Are Blind

Orientation VR systems allow their developers to integrate special action commands that are not available to people who are blind in RS. In this research, we examined if and how VR technology affects spatial and cognitive abilities. The BlindAid and Virtual Cane systems included action commands that are available to people who are blind only in the VR system. The BlindAid system includes action commands such as teleport (move the user to the starting point), additional audio information about the object, exploring the VE's structure layout without objects in it, or exploring the VE's structure with objects. The Virtual Cane system included the teleport action command and the ability to explore the VE using look-around mode. These unique capabilities supported the VE participants during the exploration process and in cognitive map construction and later assisted them in orienting themselves in the RS. The teleport action command was used by all VR users by teleporting them directly to the entrance point in an easy, short, and simple way. The look-around mode action command, used only by the Virtual Cane participants, influenced exploration, collection of spatial information, and manner of constructing a cognitive map.

By comparing the spatial tasks of the two experimental groups, we hoped to learn more about the impact of the user-interface orientation components on spatial and cognitive abilities. The Virtual Cane participants used mainly the look-around mode, which affords the user the ability to stand in one location and to scan the components of the space. To explore RS, people who are blind rely on many information units, which they need to collect in order to decide how to navigate the space. In this way, they compensate for the lack of access to distant landmarks, which have been found to be very useful to sighted people [36]. Unlike the RS, the Virtual Cane system provides this access to distant landmarks. It allows the user to collect information about an object's identity and its distance from the user without the need to walk to it. In addition to being accurate and detailed, this VE exploration created an information load [28]. Using the look-around mode affects the participants' exploration process (duration, spatial strategy, and pauses) and their cognitive map (spatial representation). In the exploration task, they needed a longer time to explore the VE and much longer pauses, compared with the BlindAid or control groups. These pauses were not technical pauses; rather, they were used to recall the spatial information, to organize, to create the relations between each component (direction and distance), and finally to construct a cognitive map. These spatial processes might affect cognitive load, exploration duration, and pauses.

The Virtual Cane participants used the walk-around mode for a short time, mainly the object-to-object strategy, as a spatial strategy. In contrast, the BlindAid and the control groups, who used only the walk-around mode, mainly used the perimeter strategy. These results mirrored practices found in O&M rehabilitation programs, which recommend the use of the perimeter strategy first, followed by grid or object-to-object strategies, in unfamiliar indoor spaces [37].

The look-around mode had one more effect on the type of spatial representation in the construction of the cognitive map. Most of the Virtual Cane participants used the map model, in contrast to the BlindAid and control group participants, who mainly used the route model or mentioned the components as a list of items. There are two ways to represent space, by using a route model or map model; most sighted people use both models. In O&M rehabilitation training, for safety reasons, the main spatial representation that is learned is the route model [37]. This approach guides the person who is blind to concentrate on his or her target path without paying attention to other components in the space that are not related to the target path; however, situations arise in which the map model is more efficient, for example, when the path is blocked and there is a need to find an alternate path.

The unique action command of the BlindAid system focuses on exploring the space in layers: the choice of exploring the VE's structure without the objects it contains or with the objects. This action command is not available for people who are blind in the RS and is not learned in an O&M rehabilitation program, but it does exist in a visual map (paper or digital). This ability to choose the layouts and the degree of spatial resolution is mainly important when we want to adjust the O&M VR system to the user. Only one participant explored the complex VE using this action command. This unique action command might especially aid people who are blind who use a guide dog. They mainly need to collect information about the structure of the space; the guide dog will direct them around obstacles in the form of objects. In contrast, people who are blind who walk with a white cane need to survey both structure and object components to be able to avoid colliding with objects. The ability to adapt the VE to the participant's needs and to his or her mobility aid is a valuable component of the VR user-interface approach. To integrate this widely, it should be integrated into O&M rehabilitation training.

4.3. Implications for Researchers and Developers

Further research should explore the integration of VR systems in O&M rehabilitation programs and the impact of the special action commands on the spatial ability of people who are blind. For example, it should evaluate the impact of the VE exploration using the look-around mode on RS orientation tasks that encompass the need to find an alternate path. It would also be useful to study differences in spatial orientation behavior of participants who use a guide dog versus a white cane and the implications of these differences for the design of a VE user interface.

Further development of the next orientation VR systems should be address three factors: intuitive and simple interfaces [25], O&M rehabilitation program theories [37], and adaptive action commands that are able to enhance spatial abilities in a manner not available in RS as discussed above.

Finally, future VR systems might be based on smartphones [38–40], wearable technologies [41,42], and crowd-sourced navigation technologies. Since smartphones, as mainstream devices, are more affordable and easier to use, most people who are blind and O&M specialists use them. For example, the X-Road project [38] uses a smartphone and a headset to allow users who are visually impaired to explore RS via a VR system. A crowd-sourced navigation app powered and monitored by the walking community, such as the Waze application, can aid people who are blind to update spatial information about their target path or area.

Funding: This research was funded by National Institutes of Health—National Eye Institute grant number 5R21EY16601-2 and by European Commission grant number FP7-PEOPLE-2007-4-3-IRG.

Institutional Review Board Statement: This study protocol was approved by the Ethics Committee of Tel Aviv University.

Informed Consent Statement: Informed consent was obtained from all participants involved in the study.

Acknowledgments: The BlindAid system was developed via a collaborative research project with my colleagues from the Massachusetts Institute of Technology (MIT): Srinivasan, M.A., Schloerb, D. and Kummar, S., and was supported by a grant from the National Institutes of Health—National Eye Institute (Grant No. 5R21EY16601-2). The Virtual Cane system was developed in collaborative research with my colleagues from Nottingham University: Brown, D.J., Battersby, S., Evett, L. and Merritt, P., and was partly supported by a grant from European Commission, Marie Curie International Reintegration Grants (Grant No. FP7-PEOPLE-2007-4-3-IRG). I thank my students Gedalevitz, H. and Milman, I. and the anonymous research participants for their time, effort, and ideas.

Conflicts of Interest: The authors declare no conflict of interest.

References

1. Jacobson, W.H. *The Art and Science of Teaching Orientation and Mobility to Persons with Visual Impairments*; American Foundation for the Blind: New York, NY, USA, 1993.
2. Roentgen, U.R.; Gelderblom, G.J.; Soede, M.; de Witte, L.P. Inventory of electronic mobility aids for persons with visual impairments: A literature review. *JVIB* **2008**, *102*, 702–724. [CrossRef]
3. Lahav, O. Improving orientation and mobility skills through virtual environments for people who are blind: Past research and future potential. *Int. J. Child Health Hum. Dev.* **2014**, *7*, 349–355.
4. Schultheis, M.T.; Rizzo, A.A. The application of virtual reality technology for rehabilitation. *Rehabil. Psychol.* **2001**, *46*, 296–311. [CrossRef]
5. Kreimeier, J.; Götzelmann, T. Two decades of touchable and walkable virtual reality for blind and visually impaired people: A high-level taxonomy. *Multimodal Technol. Interact.* **2020**, *4*, 79. [CrossRef]
6. Lahav, O.; Schloerb, D.W.; Srinivasan, M.A. Rehabilitation program integrating virtual environment to improve orientation and mobility skills for people who are blind. *Comput. Educ.* **2015**, *80*, 1–14. [CrossRef] [PubMed]
7. Evett, L.; Battersby, S.J.; Ridley, A.; Brown, D.J. An interface to virtual environments for people who are blind using Wii technology—Mental models and navigation. *JAT* **2009**, *3*, 26–34. [CrossRef]
8. Lahav, O.; Mioduser, D. Exploration of unknown spaces by people who are blind using a multi-sensory virtual environment. *JSET* **2004**, *19*, 15–23. [CrossRef]
9. Lahav, O.; Schloerb, D.; Kummar, S.; Srinivasan, M.A. A virtual map to support people who are blind to navigate through real spaces. *JSET* **2011**, *26*, 41–57.
10. Max, M.L.; Gonzalez, J.R. Blind persons navigate in virtual reality (VR): Hearing and feeling communicates "reality." In *Medicine Meets Virtual Reality: Global Healthcare Grid*; Morgan, K.S., Hoffman, H.M., Stredney, D., Weghorst, S.J., Eds.; IOS Press: San Diego, CA, USA, 1997; pp. 54–59.
11. Ohuchi, M.; Iwaya, Y.; Suzuki, Y.; Munekata, T. Cognitive-map formation of blind persons in virtual sound environment. In Proceedings of the 12th International Conference on Auditory Display, London, UK, 19–24 June 2006.
12. Pokluda, L.; Sochor, J. Spatial haptic orientation for visually impaired people. In *The Eurographics Ireland Chapter Workshop Series*; University of Ulster: Coleraine, UK, 2003.
13. Simonnet, M.; Vieilledent, S.; Jacobson, D.R.; Tisseau, J. The assessment of non-visual maritime cognitive maps of a blind sailor: A case study. *J. Maps* **2010**, *6*, 289–301. [CrossRef]
14. Torres-Gil, M.A.; Casanova-Gonzalez, O.; Gonzalez-Mora, J.L. Applications of virtual reality for visually impaired people. *WSEAS Trans. Comput.* **2010**, *2*, 184–193.
15. Merabet, L.; Sánchez, J. Development of an audio-haptic virtual interface for navigation of large-scale environments for people who are blind. Lecture notes in computer science. In In Proceedings of the 10th International Conference, UAHCI 2016, Held as Part of HCI International, Toronto, ON, Canada, 17–22 July 2016; pp. 595–606.
16. Guerreiro, J.; Sato, D.; Ahmetovic, D.; Ohn-Bar, E.; Kitani, K.M.; Asakawa, C. Virtual navigation for blind people: Transferring route knowledge to the real-World. *Int. J. Hum.-Comput. Stud.* **2020**, *135*, 102369. [CrossRef]
17. Bowman, E.L.; Liu, L. Individuals with severely impaired vision can learn useful orientation and mobility skills in virtual streets and can use them to improve real street safety. *PLoS ONE* **2017**, *12*, e0176534. [CrossRef]

18. D'Atri, E.; Medaglia, C.M.; Serbanati, A.; Ceipidor, U.B.; Panizzi, E.; D'Atri, A. A system to aid blind people in the mobility: A usability test and its results. In Proceedings of the Second International Conference on Systems, Sainte-Luce, France, 10–13 April 2007.
19. González-Mora, J.L.; Rodriguez-Hernández, A.F.; Burunat, E.; Martin, F.; Castellano, M.A. Seeing the world by hearing: Virtual acoustic space (VAS) a new space perception system for blind people. In Proceedings of the IEEE International Conference on Information and Communication Technologies, Damascus, Syria, 24–28 April 2006.
20. Seki, Y.; Ito, K. Study on acoustical training system of obstacle perception for the blind. In *Assistive Technology Research Series 11, Assistive Technology—Shaping the Future*; IOS Press: Dublin, Ireland, 2003; pp. 461–465.
21. Seki, Y.; Sato, T. A training system of orientation and mobility for blind people using acoustic virtual reality. *IEEE Trans. Neural Syst. Rehabil. Eng.* **2011**, *19*, 95–104. [CrossRef]
22. Ungar, S.; Blades, M.; Spencer, S. The construction of cognitive maps by children with visual impairments. In *The Construction of Cognitive Maps*; Portugali, J., Ed.; Kluwer Academic: Dordrecht, The Netherlands, 1996; pp. 247–273.
23. Kitchin, R.M. Cognitive maps: What are they and why study them. *J. Environ. Psychol.* **1994**, *14*, 1–19. [CrossRef]
24. Noordzij, M.L.; Zuidhoek, S.; Postma, A. The influence of visual experience on the ability to form spatial mental models based on route and survey descriptions. *Cognition* **2006**, *100*, 321–342. [CrossRef]
25. Norman, D.A. *The Design of Everyday Things*, 2nd ed.; The Perseus Books Group: New York, NY, USA, 2013.
26. Froehlich, J.; Brock, A.; Caspi, A.; Guerreiro, J.; Hara, K.; Kirkham, R.; Schöning, J.; Tannert, B. Grand challenges in accessible maps. *Univers. Interact.* **2019**, *26*, 78–81. [CrossRef]
27. Lazar, J.; Feng, J.H.; Hochheiser, H. *Human-Computer Interaction*, 2nd ed.; Elsevier, Morgan Kaufmann Publisher: Cambridge, MA, USA, 2017. Available online: https://books.google.co.il/books?hl=en&lr=&id=hbkxDQAAQBAJ&oi=fnd&pg=PP1&dq=human+computer+interface+research&ots=Sp232b235W&sig=Y758ptVO7zxAOrrc8GxWVr5uCpE&redir_esc=y#v=onepage&q=human%20computer%20interface%20research&f=true (accessed on 21 October 2021).
28. Lahav, O.; Gedalevitz, H.; Battersby, S.; Brown, D.J.; Evett, L.; Merritt, P. Virtual environment navigation with look-around mode to explore new real spaces by people who are blind. *Disabil. Rehabil.* **2018**, *40*, 1072–1084. [CrossRef] [PubMed]
29. Lahav, O. Blind Persons' Cognitive Mapping of Unknown Spaces and Acquisition of Orientation Skills, by Using Audio and Force-Feedback Virtual Environment. Unpublished Doctoral Dissertation, Tel Aviv University, Tel Aviv, Israel, 2003.
30. Schloerb, D.W.; Lahav, O.; Desloge, J.G.; Srinivasan, M.A. BlindAid: Virtual environment system for self-reliant trip planning and orientation and mobility training. In Proceedings of the IEEE Haptics Symposium, Waltham, MA, USA, 25–26 March 2010; pp. 363–370.
31. Battersby, S. The Nintendo Wii controller as an adaptive assistive device—A technical report. In Proceedings of the HEA ICS Supporting Disabled Students through Games Workshop, Middlesbrough, UK, 4 February 2008.
32. Peek, B. Managed Library for Nintendo's Wiimote. 2008. Available online: http://www.brianpeek.com (accessed on 21 October 2021).
33. Jacobson, R.D.; Kitchin, R.; Garling, T.; Golledge, R.; Blades, M. Learning a complex urban route without sight: Comparing naturalistic versus laboratory measures. In Proceedings of the International Conference of the Cognitive Science Society of Ireland, Mind III, 17–19 August 1998; University College Dublin: Dublin, Ireland.
34. Hill, E.; Rieser, J.; Hill, M.; Hill, M.; Halpin, J.; Halpin, R. How persons with visual impairments explore novel spaces: Strategies of good and poor performers. *JVIB* **1993**, *87*, 295–301. [CrossRef]
35. Hill, E.; Ponder, P. *Orientation and Mobility Techniques: A Guide for the Practitioner*; American Foundation for the Blind Press: New York, NY, USA, 1976.
36. Lynch, K. *The Image of the City*; MIT Press: Cambridge, MA, USA, 1960.
37. Campbell, N. Mapping. In *The Sound of Silence*; Rosenbaum, R., Ed.; The Carroll Center for The Blind: Newton, MA, USA, 1992.
38. Thevin, L.; Briant, C.; Brock, A.M. X-Road: Virtual reality glasses for orientation and mobility training of people with visual impairments. *ACM Trans. Access. Comput.* **2020**, *13*, 1–47. [CrossRef]
39. Cobo, A.; Serrano Olmedo, J.J.; Guerrón, N.E. Methodology for building virtual reality mobile applications for blind people on advanced visits to unknown interior spaces. In Proceedings of the 14th International Conference Mobile Learning, Lisbon, Portugal, 14–16 April 2018.
40. Guerrón, N.E.; Cobo, A.; Serrano Olmedo, J.J.; Martín, C. Sensitive interfaces for blind people in virtual visits inside unknown spaces. *Int. J. Hum.-Comput. Stud.* **2020**, *133*, 13–25. [CrossRef]
41. Siu, A.F.; Sinclair, M.; Kovacs, R.; Ofek, E.; Holz, C.; Cutrell, E. Virtual reality without vision: A haptic and auditory white cane to navigate complex virtual worlds. In *CHI 2020*; ACM Press: Honolulu, HI, USA, 2020.
42. Sinclair, M.; Ofek, E.; Gonzalez-Franco, M.; Holz, H. CapstanCrunch: A haptic VR Controller with user-supplied force feedback. In Proceedings of the 32nd Annual ACM Symposium on User Interface Software and Technology (UIST'19); Association for Computing Machinery: New York, NY, USA, 2019; pp. 815–829.

sensors

MDPI

Communication

Intuitive Cognition-Based Method for Generating Speech Using Hand Gestures

Eldad Holdengreber [1,2,*], Roi Yozevitch [3] and Vitali Khavkin [1]

1 Department of Mechanical Engineering and Mechatronics, Ariel University, Ariel 40700, Israel; vitalyk@ariel.ac.il
2 Navigation and Accessibility Research Center of Ariel University (NARCA), Ariel 40700, Israel
3 Department of Computer Science, Holon Institute of Technology, Holon 58102, Israel; roiyo@hit.ac.il
* Correspondence: eldadh@ariel.ac.il

Abstract: Muteness at its various levels is a common disability. Most of the technological solutions to the problem creates vocal speech through the transition from mute languages to vocal acoustic sounds. We present a new approach for creating speech: a technology that does not require prior knowledge of sign language. This technology is based on the most basic level of speech according to the phonetic division into vowels and consonants. The speech itself is expected to be expressed through sensing of the hand movements, as the movements are divided into three rotations: yaw, pitch, and roll. The proposed algorithm converts these rotations through programming to vowels and consonants. For the hand movement sensing, we used a depth camera and standard speakers in order to produce the sounds. The combination of the programmed depth camera and the speakers, together with the cognitive activity of the brain, is integrated into a unique speech interface. Using this interface, the user can develop speech through an intuitive cognitive process in accordance with the ongoing brain activity, similar to the natural use of the vocal cords. Based on the performance of the presented speech interface prototype, it is substantiated that the proposed device could be a solution for those suffering from speech disabilities.

Keywords: hand gestures recognition; muteness; speech disability; depth camera; Leap Motion Controller; speech interface; cognitive sensing; information entropy

1. Introduction

Muteness at its various levels is a common disability [1]. This disability can be caused due to a number of factors such as nervous system diseases, stroke, sclerosis, autism, anatomical resections, and many more. There are a number of non-surgical supportive and alternative communication technology solutions, such as eye control communication ("Hawking language"), typing, handwriting, sign language using aids, etc. In most cases, these solutions lack the flexibility to become a decent substitute for human language and in other cases require prolonged learning of a new language (e.g., sign language). Simple handwriting can be a good solution, but one that does not allow verbal communication and is less suitable in cases where visual communication is irrelevant.

In recent years, a number of applications and interfaces have been developed to help people with speech disabilities. Many of the leading methods in the field are based on hand gesture recognition. Some studies have demonstrated the identification of general physiological gestures to vocal speech [2,3], and other studies showed a direct simultaneous translation from sign language to vocal speech [4–7]. In addition to sign language interpretation, hand gesture recognition [8] is attracting a growing interest due to its applications in many different fields, such as human–computer interaction, robotics, computer gaming, and so on. 3D hand pose estimation can also be achieved via neural networks [9,10]. The Leap Motion Controller device (*LMC*), first introduced in 2012, has opened new opportunities for gesture recognition [11]. The *LMC* is essentially a depth camera device that captures the movements of the hands with high accuracy [12,13]. Its companion algorithm enables the device (according to the manufacturer) to reach an accuracy level of ≈0.01 mm. Previous studies have also suggested the *LMC* can be a useful tool for translating sign language and hand gestures into sounds [14–16].

In this study, we propose a different approach for helping those suffering from speech disabilities. The proposed technology is based on the *LMC* for hands' gesture recognition. Hand gestures are translated into vowels or consonants and those, in turn, are translated into syllables, according to the phonetic division (see Section 3 for a detailed explanation). This method relies on an intuitive approach to the vocal languages as words being comprised of syllables. The gesture-phonetic mapping goes coarsely like this: Hand gestures are divided according to different attitude (yaw, pitch, and roll) and position. Each attitude–position combination is uniquely linked to a vowel–consonant pair. A dedicated software plays this syllable in real time using speakers. There are several speech synthesis approaches and this paper restricts itself to concatenative synthesis approach [17].

Through a process of self-learning the user can intuitively communicate his/her desired "speech sounds" through the system, thus creating a working replacement for his/her dysfunctional speech system. The process itself works as follows: The user is taught the gesture–syllable map (which gesture is responsible for which syllable). Once the user gets the gist of it, we start the self-learning process. In this process, the user tries to create sounds and words using hand gestures and instant vocal feed-back from the system.

2. The Intuitive Cognition Speech Interface

In this section, we elaborate on the basic framework of the Intuitive Cognition Speech Interface (ICSP). The ICSP is comprised both from Hardware (the LMC and speakers) and software. The fusion between the two creates the speech interface.

2.1. Hardware

The *LMC* is a new consumer-grade sensor developed by Leap Motion (Leap Motion, http://www.ultraleap.com, accessed on 14 March 2021). It is primarily designed for hand gesture and finger position detection in interactive software applications (mainly gaming and 3-D CAD). Because of the current patent pending, only insufficient information on the underlying software's geometrical or mathematical frameworks is available. Figure 1 shows a schematic view of the controller's hardware setup. Differently from other devices (such as the Microsoft Kinect [18]), the *LMC* is explicitly targeted to hand gesture recognition and directly computes with high accuracy the hands' positions and postures. The following figure shows both the controller itself and the 3D output of two hands simultaneously.

Figure 1. The Leap Motion controller (LMC) device showing a 3-D representation of both hands.

The controller itself reports the position, velocity, and orientation vector of each finger. As depicted in Figure 2, the controller can detect the relative position of two (or more) hands with high precision and in 3D. The position of each hand's palm is reported relative to the sensor's center (0,0,0).

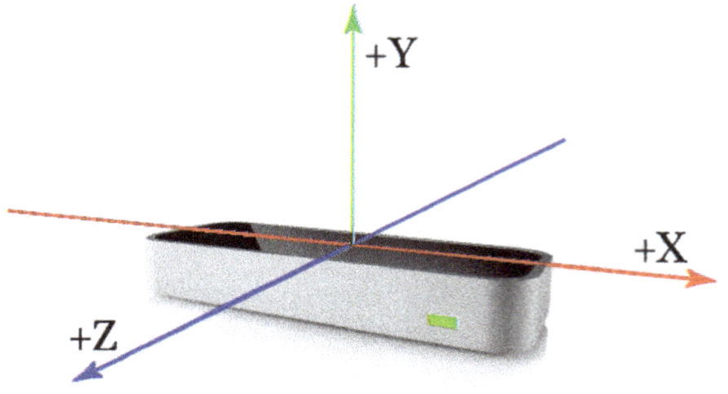

Figure 2. The LMC X-Y-X orientation.

When the right hand moves away to the right, its x value increases. When the left hand moves away to the left, its x value decreases. The Y-axis is associated with the hand's height relative to the sensor (there is no negative height). The Z-axis is associated with depth. For example, in Figure 1, when the hand approaches the screen, the Z value decreases. The LMC also provides data for hands' yaw, pitch and roll. Figure 3 describes the difference between those attitudes.

Figure 3. Hand's different attitudes.

In addition, the LMC can provide 2 more valuable pieces of information:

1. Hand Grip; it is an indication of how closed the fist is.
2. The thumb direction compared to the hand.

2.2. *Software*

The software was developed using Unity3D, a sophisticated game engine that allows for a relatively simple interaction with the *LMC*. An important feature of Unity is the ability to see in real time a 3D representation of the hands (as can be seen in Figure 1). Once the exact hands' posture is captured, the software translates it to vowels and consonant, fuses them to a distinguished sound (syllable), and plays this sound via the speakers.

3. Sound Segmentation

Words are constructed from syllables, which, in turn, are constructed from vowels and consonants. This division is relevant for non-tonal languages such as (almost) all European languages, Hebrew, Arabic etc. [19]. Therefore, the basic building blocks are those two-vowels and consonants. Let us start with vowels.

3.1. Vowels

American English has seven vowel letters (A, E, I, O, U, Y, W), but those seven letters do not encapsulate the richness of all vowel sounds. For example, the letter **a** can be sound as /**a**/ (like in the word "**father**") or as /**æ**/ (like in the word "**apple**").

Those small but important distinctions were ignored in the proposed simulator. In the final section, we suggest some methods to tackle this challenge.

Vowels are produced by different right-hand postures.

3.2. Consonants

The American English language contains 21 different consonants letters which produce 24 consonants sounds in most English accents [20]. Because of the history of the English language, there is no neat one-to-one relationship between letter and sound. Some vowels can be used also as consonants. For example, in the word *yellow*, y is a consonant, but in the word *happy*, y is a vowel.

Consonants are expressed using the left hand's posture. The combination of Left-Right hand creates syllables. The latter can be further divided to simple and advance syllables.

3.3. Simple Syllables

We define simple syllables as syllables which contains a single vowel–consonant combination. For example, the sound "B-ee" (as in the word "*be*") and the sound "A-r" (as in the word "*are*") are simple syllables. They are simple as their sound can be produced from a simple right-left hand posture combination. Figure 4 demonstrates it well. The right hand posture for the sound "*ee*" is depicted in the left side, and the left-hand posture for "*B*" is depicted in the right side image. The sound "B-ee" is simply constructed from those two postures.

Figure 4. Hands posture for a "bee" sound.

Although simple, those syllables do contain a slight complexity: the letter order. The combination R-A can be heard as "a-r" where the consonant follows the vowel (as in the word "are") or as "r-a" where the vowel follows the consonant (as in the word "e*ra*"). This distinction is solved by the vertical position (y-axis) of the right hand. Thus, the syllable "A-r" will be expressed almost identical to the syllable "R-a" with the right-hand vertical position change. Figure 5 shows this difference.

3.4. Advanced Syllables

Many words cannot be broken down to a simple syllables sequence. The words Dad, Mom, Want, Father, and Dog to name only a few, are more complex. Their complexity lies in the additional consonant (usually at the end). For example, the word "dog" is comprised of a single (complex) syllable with two consonants ("d" and "g") wrapping a vowel ("o").

These words can be expressed using the proposed system as a combination of a simple syllable and a single consonant. For example, the word "dog" can be comprised of the simple syllable "do" and the consonant "g". Although the proposed solution is not ideal (the word dog is pronounced differently than "do-g"), the word can be easily understood. Another important feature this method enables is the small numbers of hand postures to memorize. Moreover, given some knowledge about the distribution and entropy of the English language, one can improve the system even more.

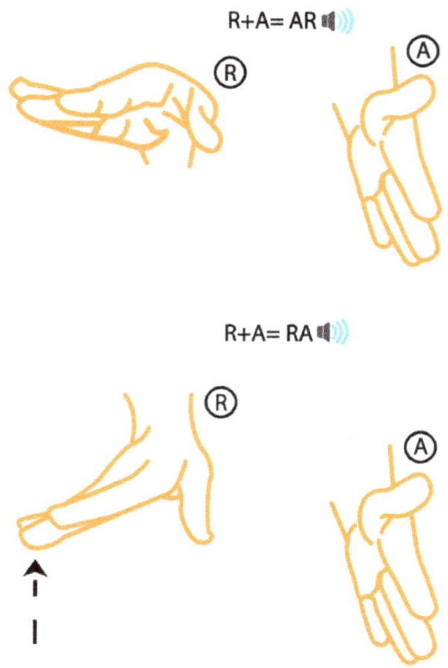

Figure 5. Hands' posture for *AR* and *RA* sounds.

4. Entropy

Information entropy is a profound concept in the field of communication. The name was coined by Claude Shannon in a pioneering paper [21]. In a nutshell, the entropy of a system can be computed as

$$H(x) = -\Sigma_{i=1}^{N} P(x_i) log_2 P(x_i) \quad (1)$$

where N is the number of different symbols, and $p(x_i)$ is the probability (or frequency) of i-th symbol. One can think of this equation as the average number of bits necessary to communicate a message. This message can represent an arbitrary distribution. When the distribution is uniform, information entropy gets its maximal value.

What is the entropy of the English alphabet? Assuming 27 letters (26 letters plus a "space" character) and a uniform distribution among all letters, the entropy is $\approx$4.75. It means that on average it would take 4.75 bits to communicate a single letter (for 32 uniform symbols, the entropy $H(x) = 5$ bits as $2^5 = 32$). This is also called the "zero-order" model of English. However, the English alphabet is not distributed uniformly. For example, the probability of the letter "e" is 100 bigger than the probability of the letter 'z'. In fact, Shannon himself found that when taking those different probabilities into account, the entropy decreases to $H(x) = 4.219$.

The above figure (4.219 bits) is also called the "first-order" model of English. The assumption of independence (zero memory) is also incorrect. Some letters follow other letters frequently; others not at all (e.g., "u" must follow "q"). One can compute the likelihood of digrams (two-letter combinations), trigrams, etc. Adding digrams to the computation gives a second-order model; adding trigrams gives a third-order model. Table 1 shows the most frequent digrams and trigrams in the English language [22].

Table 1. Most frequent digrams and trigrams in the English language.

Digrams	Trigrams
EN	ENT
RE	ION
ER	AND
NT	ING
TH	IVE
ON	TIO
IN	FOR
TR	OUR
AN	THI
OR	ONE

A "third-order model" yields 2.77 bits per symbol. The actual entropy is the "limit" of this process of taking higher and higher order models. The entropy figure is important because this paper strives to construct an alternative speech system. A key aspect in such system is to ease the pronunciation of frequent sounds ("ent"). Those frequent sound are found using entropy.

Some of the trigrams in the table are merely sounds ("ive", "ent") but some of them are also words ("and", "for", "our"). It is important to bear in mind that third-order sequences were calculated for written English, thus, the pronunciation (or sound) of these trigrams can be completely different, based on the context. For example, the trigram ION sounds different in the word "on**ion**" and in the word "celebrat**ion**" and the trigram IVE sounds different in the word "f**ive**" and in the word "effect**ive**".

5. Experiments and Results

In order to properly communicate words and phrases using the system, one must practice it thoroughly. However, this demand can be relaxed with a moderate escalation. Another important thing to bear in mind is that this paper aims solely to present a proof of concept. Past research confirms that alternative forms of communications (e.g., switching the entire visual field upside down) can become, with practice, intuitive. In order to evaluate the performance of the proposed device, a comparison was made through signal processing speech analysis between synthetic words created by the device and spoken words [23–25].

5.1. Phase 1

Phase 1 is devoted to simple words. Such words are fabricated only from unique combinations of a single vowel and a single consonant.

5.1.1. The Word "Banana"

For example, the word "Banana" can be divided into the following syllabus sounds: "Ba-Na-Na", as explained in Section 3.3. The participant is informed regrading which hands posture produces which sound and then he/she is instructed to convey simple words using the system. The conveyed word is then recorded and compared vs a normal pronunciation of the word. The left upper graph in Figure 6 shows the representation of the word "Banana" as a function of time as created using our framework. The right upper graph, demonstrates the same word as created using the human mouth. One can see that the duration of the ICSP word ($2[sec]$) is relatively long compared to that spoken by the mouth ($1.2[sec]$) [26]. The smoothing between the three syllables demonstrates the continuity of the spoken word compared to the three separated parts of the signal. Sound can also be represented as a frequency spectrum of an audio signal as it varies with time, this is called a spectrogram [27,28]. A spectrogram of sound is created from a time signal using the Fast Fourier Transform (FFT). The lower two images in the figure represent the word's spectrogram. Dark blue corresponds to low amplitudes and the brighter colors up through orange correspond to progressively stronger or louder amplitudes. Each sound is divided into short time frames of 20 ms (2000-point windows) with a 512-point overlap between successive frames. Implementing Fourier transforms on these consecutive frames can help us obtain a good approximation of frequencies across the time domain. A Hamming window is applied to each frame to significantly reduce the spectral leakage before conducting FFT. The squared-difference mean between the two signals is very low (≈0.0001), which indicates high correlation, thus, the meaning of the ICSP word can be grasped easily.

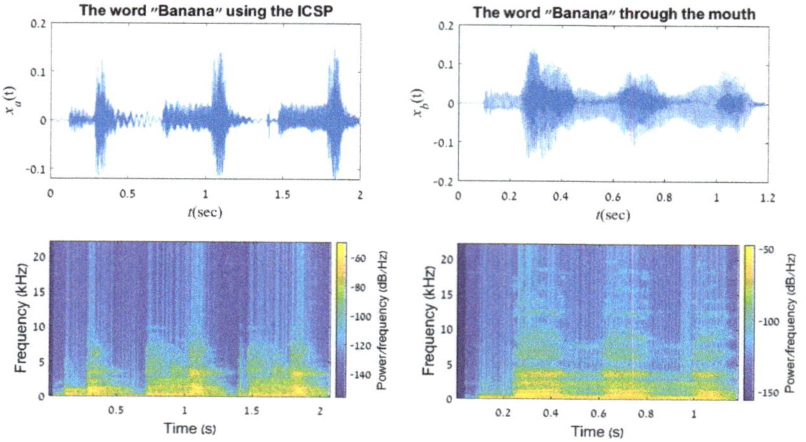

Figure 6. A representation of the word "Banana" created using the ICSP (**left graph**) and the human mouth (**right graph**). Top graphs represents the audio signals in the time domain. Down graphs are the frequency spectrum of the audio signals as it varies with time.

5.1.2. The Word "Daddy"

A sound output of the word "Daddy" can be depicted in Figure 7. The lower two images in the figure represent the word's frequency domain. Again, the figure shows similar results to Figure 6. As in the previous word, the obtained squared-difference mean is low (≈0.003). However, this figure demonstrates roughly the same duration for both words. The signal spectrogram graphs show that the two produced words contain mainly frequencies of up to 4 kHz, so that 99.9% of the signal energy is concentrated in this frequency range. The high correlation between the two demonstrates the preservation of the spectral signature of the produced word. Note that the words generated through the ICSP depends on the user level of skill, as the ability to speak through the device is an acquired skill over time. In addition, in order to get better continuity in the words spoken through the ICSP, a real-time adaptive filter can be applied for use. Another important thing to remember is that the spectral comparison serves only as a mathematical aid. The foremost goal is to create speech that can be understood by lay persons, thus, a statistical evaluation is also offered later on in Section 5.3.

5.2. Phase 2

In the second phase we introduce more complex words. Again, first the participant in informed regarding the correct posture for each sound and then instructed to convey words.

5.2.1. The Word "Thing"

As was shown before (Section 4), although the trigram "ING" is not a simple syllable, it is quite common in the English language, thus, a dedicated posture was uniquely assigned to it. The word "thing" is comprised from the sounds "thi" and "ing" in conjunction. Again, one can see the inevitable break between the two sounds on the upper left in Figure 8. The duration is ≈0.1 s and 99.5% of the signal energy is concentrated between frequency ranges up to 5 kHz.

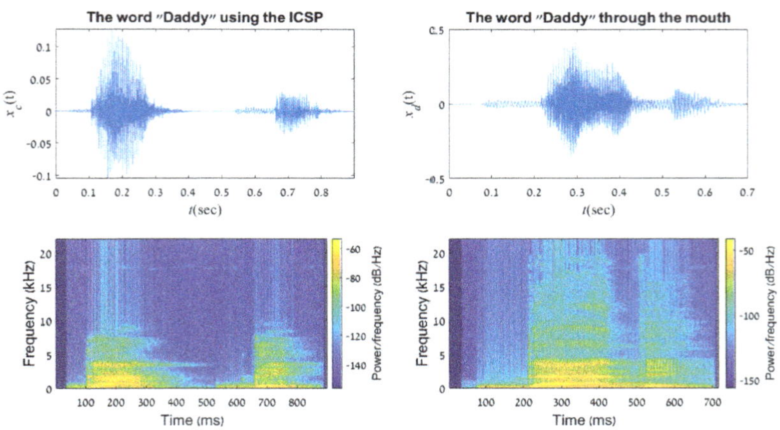

Figure 7. A representation of the word "Daddy" in time domain and frequency spectrogram, created using the ICSP (**left column**) and the human mouth (**right column**).

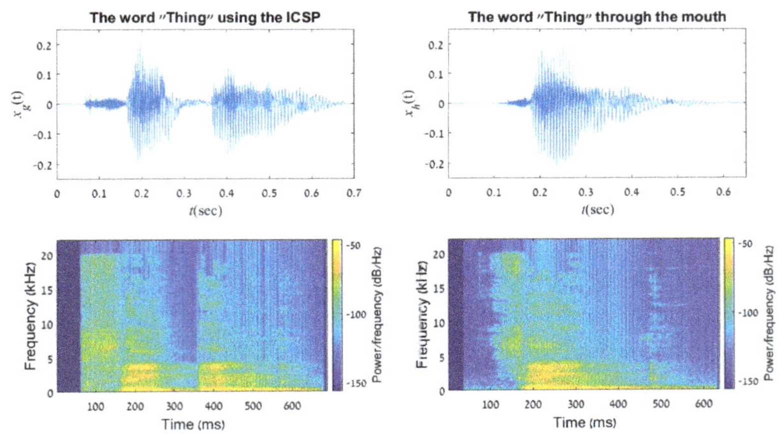

Figure 8. A representation of the word "Thing" in time domain and frequency spectrogram, created using the ICSP (**left column**) and the human mouth (**right column**).

5.2.2. The Word "Event"

The word "event", as opposed to the word "thing", is a two-syllable word (e-vent); thus, the break between the two syllables is quite similar on both words (top left and top right in Figure 9). One can find that the spectral components of those two words are with a very high match. It is indeed reflected in the ability to recognize those two words with a high degree of accuracy, as shown in the next section.

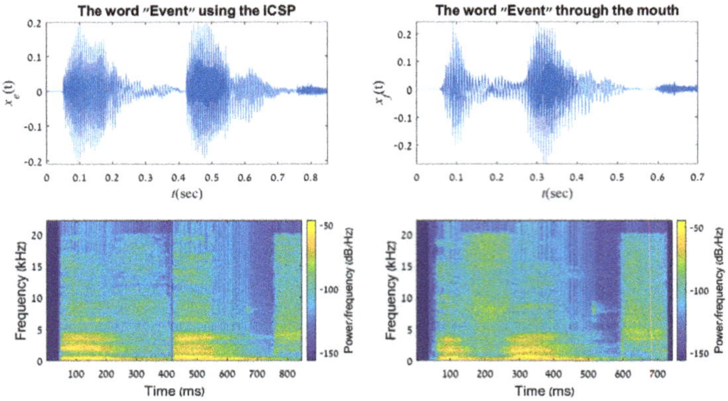

Figure 9. A representation of the word "Event" in time domain and frequency spectrogram, created using the ICSP (**left column**) and the human mouth (**right column**).

5.3. Statistic Analysis

A mandatory part of any communication protocol is the receiver's ability to understand the transmitted message. Up until this point we have focused on how to efficiently generate words using the ICSP. In this subsection, we focus on the other end of the rope: are these synthetically generated words and phrases understandable? We have generated eight different words and a one single sentence, played them to 45 mature subjects, all in the age range of 20 to 40, and checked what did the subjects hear. The subjects' native language is not English but Hebrew. The words are listed below.

- Want
- Thing
- Pizza
- Happy
- Event
- Banana
- Carrot
- Apple
- You are Happy

Table 2 shows the comprehension rate of each word/phrase. The first column represent the generated word/phrase. The second column represent the accuracy rate, i.e., the percentage of subjects who understood the correct word. For example, 53% of the subjects understood the word "want" (first word in the column). It means that 24 out of 45 subjects got this word. The two right columns represent the second best most common word and the overall accuracy of the two most common guesses. In other words, if one did not get the correct phrase, what was the second most probable phrase? In the above example, this word is "won't" which is extremely similar to the correct word. 68% of the subjects heard the word "want" or "won't". The average accuracy (first and second guesses) score for the entire set is ≈78.5%.

Table 2. Comprehension accuracy rate table.

Comprehension Accuracy Rate			
Word/Phrase	**Acc [%]**	**2nd Best Guess**	**Overall Acc [%]**
Want	53	Won't	68
Thing	24	Thinking	72
Pizza	71	Pisa	75
Happy	100	-	100
Event	82	You and	86
Banana	88	-	88
Carrot	48	Carrots	59
Apple	62	Appeal	66
You are happy	93	-	93

Although the overall accuracy rate is not perfect for some of the words ("thing"(72%), "carrot"(59%)), they can be understood from the context. Conclusions cannot be deduced from a single example but we tend to believe that the high accuracy rate for the last phrase ("you are happy") is also connected to this very point. This phenomenon is similar to automatic transcript algorithms (speech to text) that correct words backwards according to their context (just from the context, one can easily distinguish "want" from "won't").

6. Conclusions and Future Work

An intuitive cognition-based approach for playing vocal sounds through hands gestures recognition, using depth camera detection, is presented in this paper. The proposed device is capable of creating suitable vowels and consonants, consistently with a predetermined hand rotation movements. For smooth playback of the sounds and for a convenient user interface, the vocal entropy distribution and efficient software algorithmic were implemented. The prototype performances, served as a proof-of-concept, indicates that the proposed device has high potential as a solution for people with speech disabilities, while future research might show this method can bypass the traditional speech regions in the human brain. Future work might includes the aid of machine learning methods for classification and filtering [29,30].

Author Contributions: Conceptualization, V.K. and E.H.; methodology, E.H., V.K. and R.Y.; software, E.H. and R.Y.; validation, E.H. and R.Y.; formal analysis, E.H. and R.Y.; investigation, E.H. and R.Y.; resources, E.H.; writing—original draft preparation, E.H. and R.Y.; writing—review and editing, E.H. and R.Y.; visualization, E.H. and R.Y.; supervision, E.H.; project administration, E.H.; funding acquisition, E.H. All authors have read and agreed to the published version of the manuscript.

Funding: This work was supported by the Navigation and Accessibility Research Center of Ariel University (NARCA) Grant # RA2000000284.

Institutional Review Board Statement: Not applicable.

Informed Consent Statement: Not applicable.

Data Availability Statement: The data presented in this study are available on request from the corresponding author.

Conflicts of Interest: The authors declare no conflicts of interest.

References

1. Black, L.I.; Vahratian, A.; Hoffman, H.J. *Communication Disorders and Use of Intervention Services among Children Aged 3–17 Years: United States, 2012*; US Department of Health and Human Services, Centers for Disease Control and Prevention, National Center for Health Statistics: Atlanta, GA, USA, 2015.
2. Samonte, M.J.C.; Gazmin, R.A.; Soriano, J.D.S.; Valencia, M.N.O. BridgeApp: An Assistive Mobile Communication Application for the Deaf and Mute. In Proceedings of the 2019 International Conference on Information and Communication Technology Convergence (ICTC), Jeju, Korea, 16–18 October 2019; pp. 1310–1315.
3. Nagori, N.P.; Malode, V. Communication interface for deaf-mute people using microsoft kinect. In Proceedings of the 2016 International Conference on Automatic Control and Dynamic Optimization Techniques (ICACDOT), Pune, India, 9–10 September 2016; pp. 640–644.
4. Devi, S.; Deb, S. Low cost tangible glove for translating sign gestures to speech and text in Hindi language. In Proceedings of the 2017 3rd International Conference on Computational Intelligence & Communication Technology (CICT), Ghaziabad, India, 9–10 February 2017; pp. 1–5.
5. Kumar, D.N.; Madhukar, M.; Prabhakara, A.; Marathe, A.V.; Manoj; Bharadwaj, S.S. Sign Language to Speech Conversion—An Assistive System for Speech Impaired. In Proceedings of the 2019 1st International Conference on Advanced Technologies in Intelligent Control, Environment, Computing & Communication Engineering (ICATIECE), Bangalore, India, 19–20 March 2019; pp. 272–275.
6. Macriga, A.; Kruthiga, K.; Anusudha, S. Artificial Voice synthesizer using fuzzy logics. In Proceedings of the 2017 2nd International Conference on Computing and Communications Technologies (ICCCT), Chennai, India, 23–24 February 2017; pp. 388–392.
7. Nasrany, C.; Abdou, R.B.; Kassem, A.; Hamad, M. S2LV—A sign to letter and voice converter. In Proceedings of the 2015 International Conference on Advances in Biomedical Engineering (ICABME), Beirut, Lebanon, 16–18 September 2015; pp. 185–188.

8. Wachs, J.P.; Kölsch, M.; Stern, H.; Edan, Y. Vision-based hand-gesture applications. *Commun. ACM* **2011**, *54*, 60–71. [CrossRef]
9. Yang, S.; Liu, J.; Lu, S.; Er, M.H.; Kot, A.C. Collaborative learning of gesture recognition and 3D hand pose estimation with multi-order feature analysis. In *European Conference on Computer Vision*; Springer: Cham, Switzerland, 2020; pp. 769–786.
10. Zhang, F.; Bazarevsky, V.; Vakunov, A.; Tkachenka, A.; Sung, G.; Chang, C.L.; Grundmann, M. MediaPipe Hands: On-device Real-time Hand Tracking. *arXiv preprint* **2020**, arXiv:2006.10214.
11. Motion, L. Leap Motion Controller. 2015. Available online: https://www.leapmotion.com (accessed on 19 July 2021)
12. Guzsvinecz, T.; Szucs, V.; Sik-Lanyi, C. Suitability of the Kinect sensor and Leap Motion controller—A literature review. *Sensors* **2019**, *19*, 1072. [CrossRef] [PubMed]
13. Bachmann, D.; Weichert, F.; Rinkenauer, G. Evaluation of the leap motion controller as a new contact-free pointing device. *Sensors* **2015**, *15*, 214–233. [CrossRef]
14. Avola, D.; Bernardi, M.; Cinque, L.; Foresti, G.L.; Massaroni, C. Exploiting recurrent neural networks and leap motion controller for the recognition of sign language and semaphoric hand gestures. *IEEE Trans. Multimed.* **2018**, *21*, 234–245. [CrossRef]
15. Mittal, A.; Kumar, P.; Roy, P.P.; Balasubramanian, R.; Chaudhuri, B.B. A modified LSTM model for continuous sign language recognition using leap motion. *IEEE Sens. J.* **2019**, *19*, 7056–7063. [CrossRef]
16. Sharma, A.; Yadav, A.; Srivastava, S.; Gupta, R. Analysis of movement and gesture recognition using Leap Motion Controller. *Procedia Comput. Sci.* **2018**, *132*, 551–556. [CrossRef]
17. Oloko-oba, M.; TS, I.S.O. Text-to-speech synthesis using concatenative approach. *Int. J. Trend Res. Dev.* **2016**, *3*, 559–462.
18. Zhang, Z. Microsoft kinect sensor and its effect. *IEEE Multimed.* **2012**, *19*, 4–10. [CrossRef]
19. Duanmu, S. Tone and non-tone languages: An alternative to language typology and parameters. *Lang. Linguist.* **2004**, *5*, 891–923.
20. Crystal, D. *The Cambridge Encyclopedia of the English Language*; Ernst Klett Sprachen: Stuttgart, Germany, 2004.
21. Shannon, C.E. *The Mathematical Theory of Communication, by CE Shannon (and Recent Contributions to the Mathematical Theory of Communication), W. Weaver*; University of Illinois Press: Champaign, IL, USA, 1949.
22. Young, D.B. Foundations of Computer Security. Available online: https://www.cs.utexas.edu/~byoung/cs361/lecture35.pdf (accessed on 19 July 2021).
23. Lynn, P.A. *An Introduction to the Analysis and Processing of Signals*; Macmillan International Higher Education: London, UK, 2016.
24. Vargas, J.; McLaughlin, S. Speech analysis and synthesis based on dynamic modes. *IEEE Trans. Audio Speech Lang. Process.* **2011**, *19*, 2566–2578. [CrossRef]
25. Van Hengel, P.W.; Krijnders, J.D. A comparison of spectro-temporal representations of audio signals. *IEEE/ACM Trans. Audio Speech Lang. Process.* **2013**, *22*, 303–313. [CrossRef]
26. Paliwal, K.; Wójcicki, K. Effect of analysis window duration on speech intelligibility. *IEEE Signal Process. Lett.* **2008**, *15*, 785–788. [CrossRef]
27. Tran, T.; Lundgren, J. Drill Fault Diagnosis Based on the Scalogram and Mel Spectrogram of Sound Signals Using Artificial Intelligence. *IEEE Access* **2020**, *8*, 203655–203666. [CrossRef]
28. Xie, J.; Hu, K.; Zhu, M.; Yu, J.; Zhu, Q. Investigation of different CNN-based models for improved bird sound classification. *IEEE Access* **2019**, *7*, 175353–175361. [CrossRef]
29. Xia, S.; Wang, G.; Chen, Z.; Duan, Y.; Liu, Q. Complete random forest based class noise filtering learning for improving the generalizability of classifiers. *IEEE Trans. Knowl. Data Eng.* **2019**, *31*, 2063–2078. [CrossRef]
30. Yozevitch, R.; Ben-Moshe, B. Advanced particle filter methods. In *Heuristics and Hyper-Heuristics-Principles and Applications*; InTech: Rjeka, Croatia, 2017.

sensors

MDPI

Article

Situational Awareness: The Effect of Stimulus Type and Hearing Protection on Sound Localization

Leah Fostick * and Nir Fink

Department of Communication Disorders, Ariel University, Ariel 40700, Israel; nirfi@ariel.ac.il
* Correspondence: leah.fostick@ariel.ac.il; Tel.: +972-544925354

Abstract: The purpose of the current study was to test sound localization of a spoken word, rarely studied in the context of localization, compared to pink noise and a gunshot, while taking into account the source position and the effect of different hearing protection devices (HPDs) used by the listener. Ninety participants were divided into three groups using different HPDs. Participants were tested twice, under with- and no-HPD conditions, and were requested to localize the different stimuli that were delivered from one of eight speakers evenly distributed around them (starting from 22.5°). Localization of the word stimulus was more difficult than that of the other stimuli. HPD usage resulted in a larger mean root-mean-square error (RMSE) and increased mirror image reversal errors for all stimuli. In addition, HPD usage increased the mean RMSE and mirror image reversal errors for stimuli delivered from the front and back, more than for stimuli delivered from the left and right. HPDs affect localization, both due to attenuation and to limitation of pinnae cues when using earmuffs. Difficulty localizing the spoken word should be considered when assessing auditory functionality and should be further investigated to include HPDs with different attenuation spectra and levels, and to further types of speech stimuli.

Keywords: localization; speech; mirror image reversal errors

1. Introduction

The ability to localize sound is important for survival, originating from the need to identify potential risks in the environment, whether a stalking predator or an approaching car when crossing a street. Sound localization is derived mostly from a comparison of information perceived by both ears, such as differences in the precise time at which the sound reaches each ear (interaural time difference, ITD), and the intensity of the sound reaching each ear (interaural level difference, ILD) [1]. While ITD is more effective for localizing low frequencies (lower than 1500 Hz), and ILD is more effective for localizing high frequencies (higher than 1500 Hz), frequencies in the range of 2000–4000 Hz are poorly localized [2,3]. In addition to differences between ears, information is also obtained about the location of a sound from spectral changes that occur when it encounters body parts such as the torso, head, and pinnae [1,4]. The accuracy in localizing sound depends on the characteristics of the environment, the listener, and the stimulus.

In environmental characteristics, one basic feature among many is the spatial angle from which an acoustic stimulus is delivered relative to the listener's forehead or relative to another point of reference. In this study, we focused on acoustic stimuli presented in the horizontal plane, thus referring to horizontal positions (left, right, front, back, or any other azimuth) in reference to the forehead of a listener who is seated in the center and surrounded by equidistant sound sources. Studies have shown that the localization of stimuli delivered from the front, left, or right (relative to the forehead direction) is more accurate than those delivered from the back [5–7]. Confusing a right-delivered stimulus for a left one, or vice versa (termed here as right/left or left/right errors), reflects a difficulty in perceiving cues related to interaural differences. Confusing a front-delivered stimulus for a back one, or vice versa (termed here as front/back or back/front errors), reflects a difficulty in perceiving spectral cues, related to changes in the sound's spectra due to the pinnae, head, and torso [5,6]. Spectral cues contribute less than interaural difference cues to localization [1,8], but the localization of sound sources from the front or the back is especially difficult when some spectral cues are unavailable, such as when the pinnae are covered by earphones [4–9]. Thus, when considering the ability to localize sound, one should consider the source's azimuth in regard to the listener's forehead direction. The localization can be evaluated either by specifying the sound source azimuth in units of degrees, or by a binary generalization of the sound source hemifield. The latter reflects whether a stimulus originating from one of the hemifields would be perceived correctly or incorrectly as originating from the opposite hemifield.

The listeners' hearing ability is another factor affecting sound localization. The contribution of hearing ability to localize sound is usually evaluated among hearing-impaired listeners e.g., [10–12] or among typically hearing listeners using hearing protection devices (HPDs) e.g., [4–8,13–15]. The present study was of the latter type. HPDs affect the listener's ability to localize sound in two fundamental ways. First, they *attenuate* sound intensity, causing the sound to be less clear and distinctive, thus having a detrimental effect on the localization of the sound source [4–8,13–15]. This underscores the importance of considering the advantages and disadvantages of using HPDs with different attenuation levels [16]. Second, as mentioned earlier, some HPDs (e.g., earmuffs) cover the pinnae, thereby *reducing spectral cues* important for localizing sources emanating from in front of or behind the listener. Earmuffs therefore increase front/back or back/front confusion in localization [4,6,8]. Accordingly, when considering the effect of HPDs on localization, it is important to consider not only their attenuation, but also their type—whether in-the-ear (insert earplugs), over-the-ear (earmuffs), or a combination (double hearing protection), as the latter attributes to a larger attenuation and reduction in spectral cues. Therefore, in the present study, we compared the effects of in-the-ear, over-the-ear, and double-protection HPDs on sound localization.

Lastly, localization also depends on sound characteristics: level (amplitude), duration, and frequency (spectrum). The sound level affects localization ability by causing the sound to be more or less clear and distinctive, as discussed earlier in the context of HPDs. The sound duration affects localization accuracy [8,17] as a longer duration provides greater temporal information and energy than sounds of a shorter duration. In addition, long-duration sounds (more than ~1.5 s) provide the listener with sufficient time to move the head in the direction of the sound, thereby improving localization, particularly by reducing front/back or back/front confusion [1,8,18]. The sound spectrum affects localization accuracy via the frequency range (low-, high-, or middle-range frequencies) and bandwidth (narrow- vs. broad-band spectrum). Low frequencies are localized more accurately than high frequencies [3,19] but high frequencies increase the localization accuracy of sounds presented from the back of the listener [20,21]. In general, broad-band sounds are easier to localize than narrow-band sounds are (assuming equal energy in both), and pure tones are the most difficult to localize [3,7,22–25].

In everyday life settings, the ability to identify the location of a person addressing us is critical. In such settings, the sound stimulus to be localized will most likely be either a single word (short-duration stimulus) or a sentence (long-duration stimulus). Yet, the localization of words has rarely been studied [23–25], and has mainly been explored in the context of hearing aids and cochlear implants [11,12]. In contrast, pink noise e.g., [8,11,26], pure tones, and other narrow- or broad-band noise stimuli [3,4,6,8,10,15,17,25,27–30] have been repeatedly tested in localization studies, and studies relating to military and sporting activities (such as hunting) specifically focused on the localization of gunshots [7,31–34]. The few studies that tested the ability to localize different speech stimuli showed that speech stimuli characterized by a narrower bandwidth than pink noise or white noise were poorly localized in comparison, yet better localized than pure tones [11,23–25]. However, these few studies did not examine the effect of the environment on word localization, such as the azimuth of the sound relative to the listener, nor the listener's hearing ability.

Therefore, the aim of the current study was to test the ability to localize a monosyllabic word stimulus compared to two other stimuli with different spectra that have been frequently studied in previous localization studies, namely, pink noise and a gunshot sound. The localization of these stimuli was also tested in relation to the azimuth of the sound source and with the participants using various HPDs, thus manipulating hearing ability and spectral cues.

2. Materials and Methods

2.1. Participants

A total of 90 individuals participated in the study, aged 20–35 years (58% females). The participants were undergraduate students, recruited using advertisements within campus social networks, and screened for normal hearing (hearing thresholds ≤25 dB HL in frequencies of 500, 1000, 2000, 3000, 4000, 6000, and 8000 Hz). Exclusion criteria also included a diagnosis of a learning disability or attention deficit hyperactivity disorder. Participants were randomly divided into three groups of 30 participants each, in order to measure the effect of the three HPDs.

2.2. Stimuli

Three stimuli were compared in the current study: the Hebrew word 'esh' (fire) spoken by a male speaker, pink noise, and a single M16 assault rifle shot recorded at a distance of 200 feet (60.96 m) from the shooter. The word was recorded in a sound-treated booth, using an Electro-Voice™ RE320 microphone connected to a Focusrite™ Saffire Pro™ 24 DSP sound card, with a Hewlett Packard® computer running MAGIX® Samplitude® Pro X software. The M16 assault rifle shot was recorded by Sintonizar Productions [35]. Pink noise was generated using the Sound Forge™ Pro version 11 software of MAGIX®. The duration of the three stimuli was 409 ms for the word, 212 ms for the pink noise, and 202 ms for the gunshot with an additional reverberation tail of 800 ms. For determining the spectrum of each stimulus in the listener's ear, a GRAS™ 45CB Acoustic Test Fixture (ATF) was placed in the same setting as the head of the listener in the experiment. Stimuli administered from the speaker at an azimuth of 0° were recorded and analyzed using SINUS™ SAMURAI™ version 3.0.2® software. The data are presented in Figure 1a, with the word having the narrowest spectrum and pink noise the widest. Figure 1b presents the accumulative energy for each stimulus, with the word having the least accumulated energy over time, and the gunshot the most.

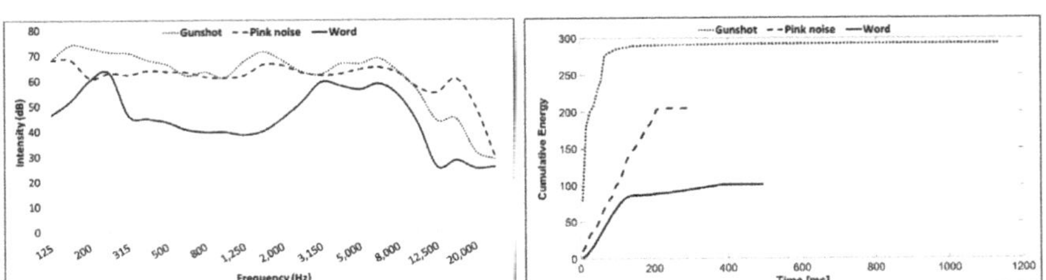

(**a**) Spectrum (**b**) Accumulative energy

Figure 1. Description of Hebrew word /esh/ (fire), pink noise, and M16 single gunshot at a distance of 200 feet (60.96 m).

2.3. Hearing Protection Devices

Three HPD conditions were tested in the current study. The HPDs were both manufactured by 3M™: Combat Arms™ 4.1 earplugs in an open mode were used for the in-the-ear condition, 3M™ Peltor™ Bull's Eye™ H515FB flat earmuffs were used for over-the-ear condition, and the combination was used for the double-protection condition.

According to the manufacturer, the Noise Reduction Rating (NRR) is 7 dB for the Combat Arms™ earplugs in open mode and is 27 dB for the Peltor™ H515FB earmuffs. Figure 2 shows the mean attenuation of each stimulus according to the various HPD configurations used in the study, as recorded by the ATF. The mean attenuation for each stimulus and HPD was calculated by subtracting the mean SPL measured for the unoccluded condition from the mean SPL measured for the occluded condition. As expected, double protection added an attenuation of at least 3 dB to single protection conditions.

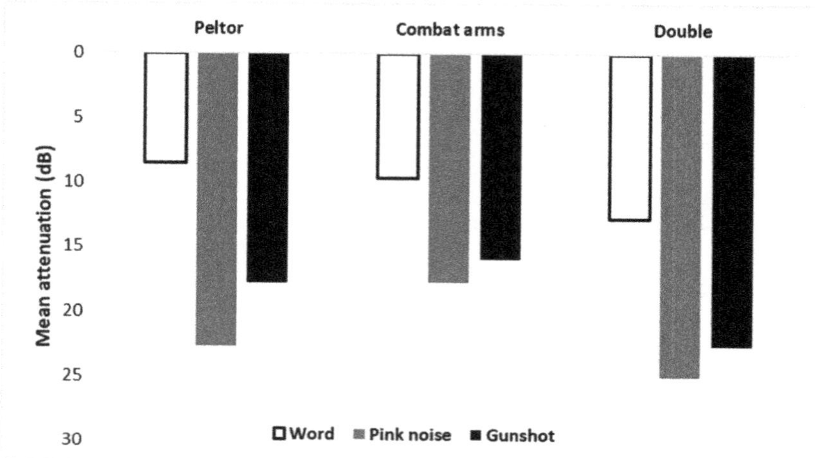

Figure 2. Mean attenuation of word, pink noise, and gunshot by HPD condition (Combat Arms™ earplugs, Peltor™ earmuffs, and double protection).

2.4. Apparatus

The experiment was carried out using a Dell™ Inspiron™ 13 5378 i5 laptop computer with designated software that controlled sound delivery and recorded participant responses. The software was written in C# version NET 4.5.2. The sounds were delivered from the computer through a Steinberg™ UR824 USB 2.0 Audio Interface into eight RCF Ayra 5® active monitors, positioned 60 cm from participants and separated by 45° starting from 22.5° through 337.5° (Figure 3a). The experimental setup was calibrated by injecting each monitor separately with a 1 kHz tone and measuring 100 dB SPL with the GRAS™ 45CB ATF, which was designed and specified to comply with the ANSI/ASA S12.42 standard. The ATF was seated in the center of the monitors' circle, resembling the position of a human participant in this experiment. The calibration tone signal from each monitor was recorded with two 1/2" microphones situated at the end of the ATF's simulated ear canals (the position of the eardrum in a human) that were connected to a SINUS™ SAMURAI™ sound level meter conforming to IEC 60651/IEC 60804/IEC 61672-1, IEC 651, and IEC 804 standards.

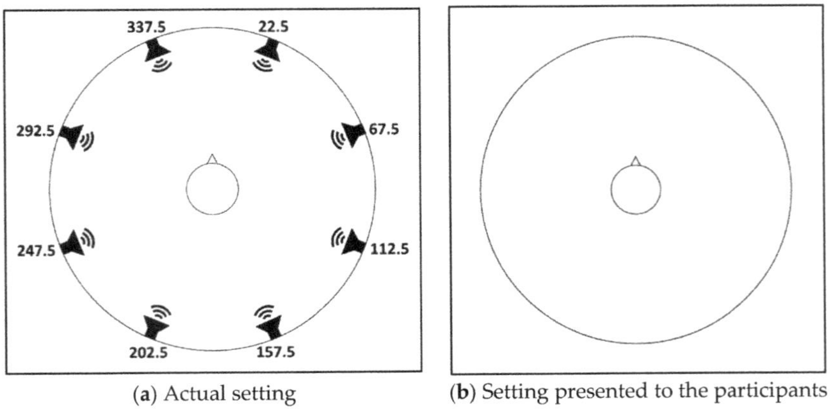

Figure 3. Experimental setting. The symbol in the middle represents participant orientation. (**a**) The actual setting: Participants sat in the middle of a circle of eight monitors separated by 45°, starting from 22.5° through 337.5°. (**b**) The setting presented to participants on a computer screen: participants were asked to indicate on the circle the location of the perceived sound source.

2.5. Procedure

The study was carried out according to Good Clinical Practice (GCP) regulations and was approved by the university institutional review board. The experiment was conducted in a sound-proof anechoic chamber. The participants sat with the computer in a tablet position on their lap. The tablet's screen presented a top-of-the-head with a tip-of-the-nose symbol representing the participant and their orientation. The participant symbol was surrounded by a circle of continuous circumference that did not indicate the monitor positions (Figure 3b). Participants were asked to respond to each stimulus by indicating the perceived location of each sound source on the circle's circumference. Each stimulus was delivered 10 times from each monitor, resulting in 240 trials (3 stimuli type × 8 monitors × 10 repetitions) randomly intermixed by the experimental software. After every 48 trials, the participants were offered a short break.

Following the provision of signed informed consent and completing the screening procedure, the participants had a short training session in which all sounds were randomly delivered once from each monitor. A break was offered after training and prior to the beginning of the experiment. Half of the participants were tested first with the HPD condition and then with the no-HPD condition; the other half was tested in the opposite order. The HPDs were fitted by the experimenter. The live experiment duration lasted approximately 20 min for both conditions, with the entire procedure (including screening and training) lasting almost 60 min. Upon completion of the task, participants received monetary compensation equivalent to USD 65 for their time.

2.6. Data Analysis

Localization accuracy was analyzed in two ways. First, we analyzed the *discriminating accuracy* of the specific perceived location of each sound source. This was analyzed in terms of the root-mean-square error (RMSE), i.e., the RMSE of the angular distance between the response angle and the target monitor angle. Second, we analyzed *mirror image reversal errors* (see also [9]); this analysis showed the degree to which participants had difficulty discriminating between sound sources due to mirror image reversal errors. Thus, the analysis focused on the percent of errors participants made when they localized the stimuli to the hemifield opposite the target monitors. Therefore, when the target monitors were on the right side (i.e., 67.5° and 112.5°), Right/Left (R/L) mirror image reversal errors were defined as responses of angles 181° to 359°, and when the target monitors were on the left side (i.e., 247.5° and 292.5°), Left/Right (L/R) mirror image reversal errors were defined as responses of angles 1° to 179°. When target monitors were in the front (i.e., 337.5° and 22.5°), Front/Back (F/B) mirror image reversal errors were defined as responses of angles 91° to 269°, and when target monitors were in the back (i.e., 157.5° and 202.5°), Back/Front (B/F) mirror image reversal errors were defined as responses of angles 271° to 360° and 0° to 89° (Figure 4).

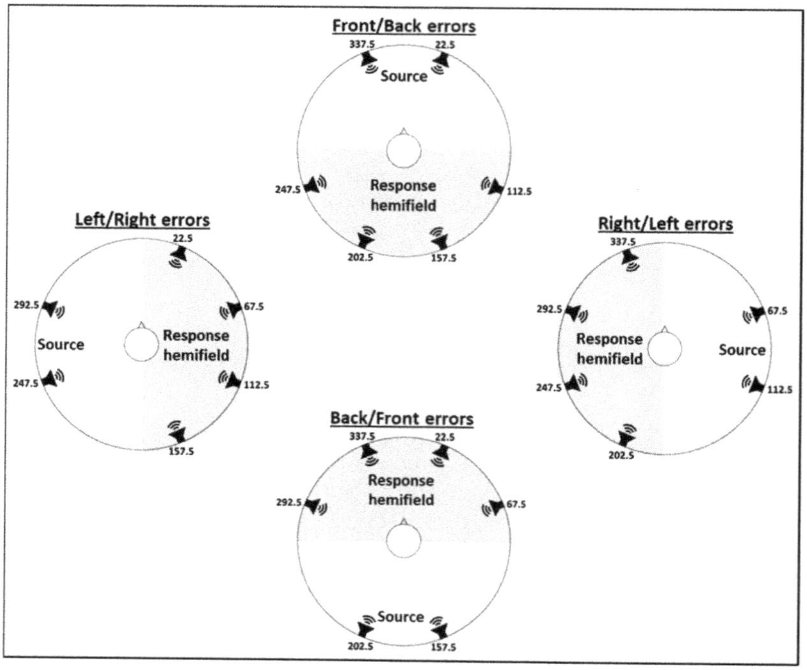

Figure 4. Mirror image reversal errors for each of the four hemifields. Errors were defined by a response opposite the hemifield of the source monitor, as indicated by the shaded areas in each hemifield.

Repeated measures ANOVAs were performed with stimuli type (word, pink noise, and gunshot) and hemifield (front, back, right, and left for hemifield discrimination) or source angle (22.5°, 67.5°, 112.5°, 157.5°, 202.5°, 247.5°, 292.5°, and 337.5° for discrimination accuracy) as within-subjects variables, and HPD type (Combat Arms™, Peltor™ H515FB, and double protection) as a between-subjects variable, on (1) discrimination accuracy (i.e., mean RMSE) and (2) mirror image reversal errors (i.e., percentage of localization to the opposite hemifield). Additional repeated measures ANOVAs and one-way ANOVAs were used when interactions were found significant. Post hoc analyses were performed using Least Significant Difference (LSD) tests. As HPDs significantly decrease localization accuracy, we chose not to compare performance with an HPD to the baseline (no-HPD), but to present the data separately, by comparing the relative baseline performance across different stimuli and, independently, comparing performance when utilizing various HPDs.

3. Results

3.1. Discrimination Accuracy

3.1.1. No-HPD (Baseline)

Main effects for RMSE were found for both stimulus type and source angle ($F(2,172) = 13.304$, $p = 0.000$, partial $\eta^2 = 0.134$; and $F(7,602) = 19.564$, $p = 0.000$, partial $\eta^2 = 0.185$, respectively). As expected, no difference in RMSE was found between HPD groups under the no-HPD condition ($F(2,86) = 2.097$, $p = 0.129$, partial $\eta^2 = 0.047$). The mean RMSE differed between all three stimuli, with the word having the largest RMSE, and the gunshot the smallest (Figure 5a). The results for source angles showed that, in general, stimuli originating from monitors located in the back (157.5° and 202.5°) had the largest RMSE, the front monitors (22.5° and 337.5°) had the smallest, and no difference in the mean RMSE was found between most of the stimuli originating from right (67.5° and 112.5°) and left (247.5° and 292.5°) monitors (Figure 5b, overall bar, and Table 1a). Interestingly, in almost all hemifields, there was a significant difference in RMSE between the two monitors positioned in the hemifield. For instance, in both the front and back hemifields, the right monitors (22.5° and 157.5°) had a larger RMSE than the left monitors did (337.5° and 202.5°), while in the left hemifield, the rear monitor (247.5°) had a larger RMSE than the front monitor did (292.5°). No difference in RMSE, however, was found between the monitors in the right hemifield (67.5°and 112.5°).

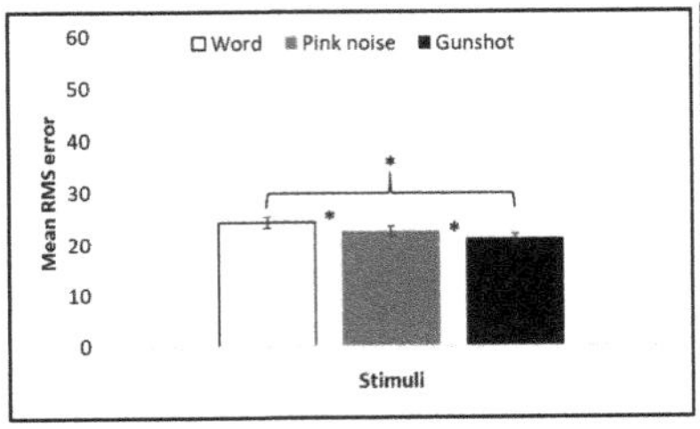

(**a**) Stimuli, no-HPD condition

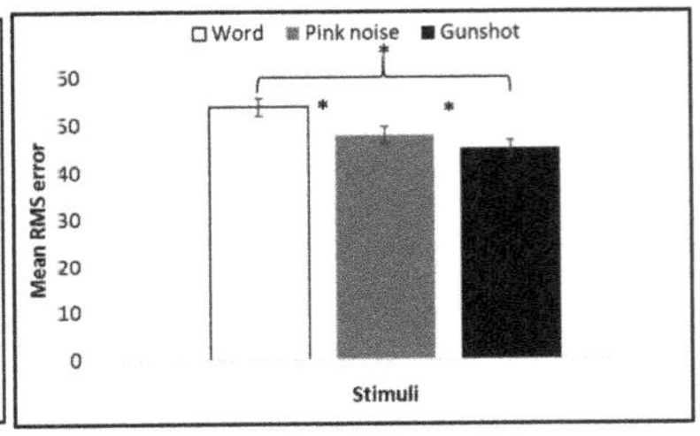

(**c**) Stimuli, HPD condition

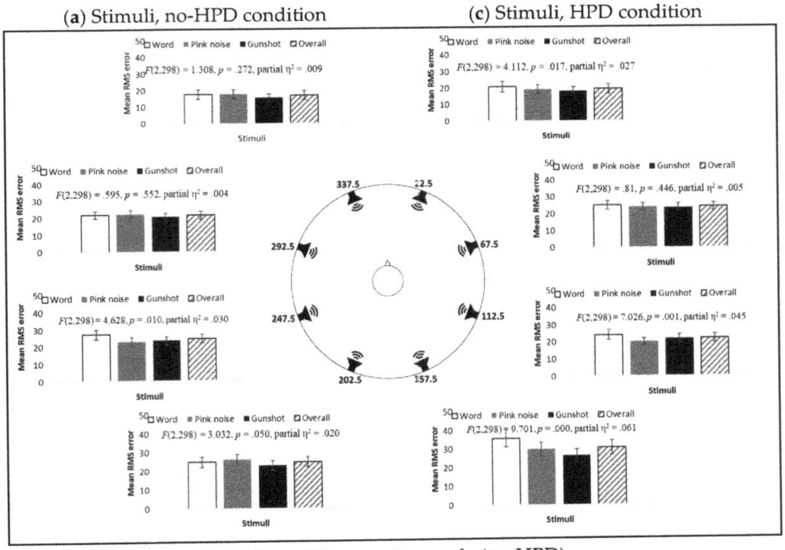

(**b**) Stimuli by monitor angle (no-HPD)

Figure 5. *Cont.*

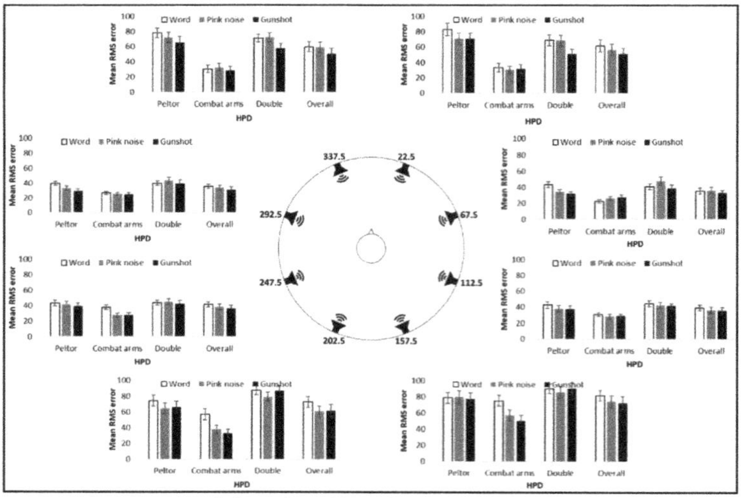

(**d**) Stimuli and HPD by monitor angle

Figure 5. Mean RMSE of stimuli for (**a**) stimuli only (no-HPD condition), averaged across all 90 participants; (**b**) stimuli only (no-HPD condition) at each source angle, averaged across all 90 participants, including statistics for main effect for stimuli at each source angle; (**c**) stimuli only (HPD condition), averaged across HPDs for all 90 participants; and (**d**) stimuli and HPD at each source angle, averaged across 30 participants in each HPD group. * $p < 0.05$.

Table 1. Results of Least-Significant-Difference (LSD) test for post hoc comparisons between the mean RMSE for monitors in different angles in the (a) non-HPD condition and (b) HPD condition (monitor angles correspond to Figure 1).

(a) no-HPD

	Back	Left		Front		Right		Back
	202.5°	247.5°	292.5°	337.5°	22.5°	67.5°	112.5°	157.5°
202.5°		0.124	3.306	7.777	5.484	0.514	2.556	5.596
247.5°			3.182	7.653	5.360	0.390	2.432	5.720
292.5°				4.471	2.178	2.792	0.750	8.902
337.5°					2.294	7.264	5.221	13.37
22.5°						4.970	2.927	11.080
67.5°							2.043	6.110
112.5°								8.152

(b) with-HPD

		157.5°	112.5°	67.5°	22.5°	337.5°	292.5°	247.5°
Back	157.5°							
Right	112.5°	39.025						
Right	67.5°	41.47	2.449					
Front	22.5°	19.78	19.24	21.69				
Front	337.5°	19.80	19.23	21.68	0.016			
Left	292.5°	42.51	3.487	1.038	22.730	22.71		
Left	247.5°	37.49	1.540	3.989	17.70	17.69	5.027	
Back	202.5°	11.04	27.99	30.434	8.742	8.758	31.473	26.446

Significant comparisons are marked with shaded cells.

A significant Stimulus Type X Source Angle interaction was found ($F(14,204) = 1.976$, $p = 0.017$, partial $\eta^2 = 0.022$). Separate repeated measures ANOVAs for each angle showed a significant main effect for stimulus type for only four of the eight source angles (Figure 5b). For all angles, the word had the largest RMSE and the gunshot had the smallest. Pink noise had a smaller RMSE than words did for source angles 112.5° and 247.5° and a larger RMSE than the gunshot did for source angle 157.5° (Table 2).

Table 2. Post hoc LSD analyses between stimuli used in the study at each monitor angle.

		Word vs. Pink Noise	Word vs. Gunshot	Pink Noise vs. Gunshot
LSD (mean RMS error, no-HPD)				
Back	157.5°	3.482	7.678	4.196
Right	112.5°	4.014	3.697	−0.317
	67.5°	1.020	1.278	0.258
Front	22.5°	1.020	3.652	0.258
	337.5°	0.085	1.693	1.608
Left	292.5°	0.087	1.101	1.014
	247.5°	3.224	3.418	0.194
Back	202.5°	−0.596	2.386	2.983

Significant comparisons are marked with shaded cells.

3.1.2. With-HPD

Main effects were found for stimulus type, source angle, and HPD type ($F(2,174) = 29.082$, $p = 0.000$, partial $\eta^2 = 0.251$; $F(2,174) = 40.580$, $p = 0.000$, partial $\eta^2 = 0.318$; and $F(2,87) = 37.222$, $p = 0.000$, partial $\eta^2 = 0.461$, respectively). As was observed in the no-HPD results, the RMSE differed between all three stimuli: the word had the largest RMSE and gunshot the smallest (Figure 5c). The mean RMSE for each stimulus at each source angle under the HPD condition, and the overall average across HPDs, are presented in Figure 5d. Post hoc LSD test results between source angles are presented in Table 1b. Similar to the results for the no-HPD condition, the largest RMSEs were found for monitors in the back (157.5° and 202.5°), although the back-left monitor (202.5°) had a smaller RMSE than the back-right one (157.5°) and a similar RMSE to the front monitors (22.5° and 337.5°). Under the HPD condition, the smallest mean RMSEs were observed in the right and left monitors (67.5°, 112.5°, 247.5°, and 292.5°). Combat Arms™ 4.1 resulted in the smallest overall mean RMSE (compared to Peltor™ H515FB: LSD = −21.009, $p = 0.000$; and compared to double protection: LSD = −24.422, $p = 0.000$). Importantly, no difference in the overall mean RMSE was found between Peltor™ H515FB and double protection (LSD = 3.413, $p = 0.269$).

Significant interactions were found for Stimulus Type X Source Angle, Source Angle X HPD Type, and Stimulus Type X Source Angle X HPD Type ($F(14,1,218) = 2.010$, $p = 0.014$, partial $\eta^2 = 0.023$; $F(14,609) = 3.587$, $p < 0.001$, partial $\eta^2 = 0.076$; and $F(28,1,218) = 2.560$, $p = 0.000$, partial $\eta^2 = 0.056$, respectively). Separate repeated measures ANOVAs for each HPD at different source angles showed that the difference between stimuli is sensitive to HPD type and source angle (Table 3): for the two front monitors (22.5° and 337.5°), a main effect for stimulus type was found only in double protection, while for the two back monitors (157.5° and 202°), along with 247.5°, a main effect for stimulus type was found only for Combat Arms™ 4.1. Monitors at angles 67.5° and 292.5° displayed a main effect for stimulus type for Peltor™ H515FB. For all these effects, the word had a larger RMSE than the gunshot. In the front monitors, pink noise also had a larger RMSE error than the gunshot (Table 4).

Table 3. Main effects for comparison between stimuli using different HPDs and from different source angles.

		Combat Arms™ 4.1		Peltor™ H515FB		Double Protection	
		F (2,58)	Partial η^2	F (2,58)	Partial η^2	F (2,58)	Partial η^2
Back	157.5°	10.465	0.265	0.078	0.003	0.421	0.014
Right	112.5°	0.461	0.016	1.089	0.036	0.294	0.010
Right	67.5°	1.717	0.056	6.947	0.193	3.433	0.106
Front	22.5°	0.357	0.012	2.770	0.087	5.585	0.161
Front	337.5°	0.581	0.020	3.068	0.096	6.104	0.174
Left	292.5°	0.617	0.021	4.668	0.139	0.707	0.024
Left	247.5°	14.405	0.332	1.098	0.036	0.256	0.009
Back	202.5°	10.346	0.263	2.001	0.065	1.738	0.057

Table 4. Post hoc analysis (LSD) for comparison between stimuli using different HPDs and from different source angles.

		Combat Arms™ 4.1			Peltor™ H515FB			Double Protection		
		Word vs. Pink Noise	Word vs. Gunshot	Pink Noise vs. Gunshot	Word vs. Pink Noise	Word vs. Gunshot	Pink Noise vs. Gunshot	Word vs. Pink Noise	Word vs. Gunshot	Pink Noise vs. Gunshot
Back	157.5°	17.744	24.854	7.110	−1.125	1.124	2.249	4.636	0.418	−4.219
Right	112.5°	2.108	1.834	−0.274	4.634	5.376	0.742	2.147	2.815	0.668
Right	67.5°	−3.403	−4.789	−1.386	9.263	11.597	2.334	−7.114	2.182	9.297
Front	22.5°	2.962	2.063	−0.899	12.268	12.115	−0.153	0.538	17.849	17.312
Front	337.5°	−2.300	1.817	4.117	5.659	12.485	6.826	−1.102	13.360	14.462
Left	292.5°	1.400	2.282	0.882	6.486	10.542	4.055	−3.730	−0.010	3.720
Left	247.5°	9.916	10.089	0.173	1.456	4.119	2.662	−1.316	1.340	2.656
Back	202.5°	18.819	23.774	4.955	9.445	7.831	−1.614	8.445	0.983	−7.462

Note. Significant comparisons are marked with shaded cells.

3.2. Mirror Image Reversal Errors

3.2.1. No-HPD (Baseline)

Main effects were found for stimulus type and hemifield, but not for HPD groups ($F(2,174) = 7.442$, $p = 0.001$, partial $\eta^2 = 0.079$; $F(3,174) = 5.823$, $p = 0.001$, partial $\eta^2 = 0.063$; and $F(2,87) = 2.274$, $p = 0.109$, partial $\eta^2 = 0.050$). The word had the largest percentage of mirror image reversal errors (Figure 6a). In addition, there were more F/B and B/F errors than R/L and L/R errors (Figure 6b, overall bar, and Table 5a).

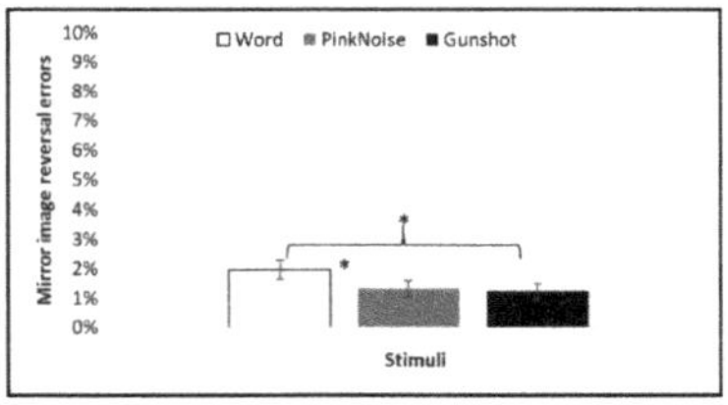

(**a**) Stimuli, no-HPD condition

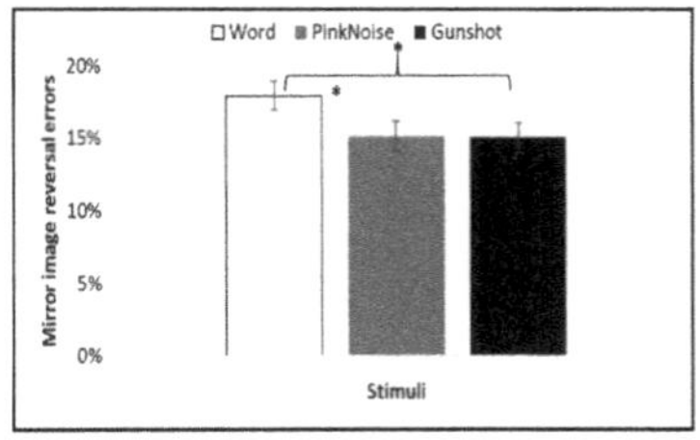

(**c**) Stimuli, with-HPD condition

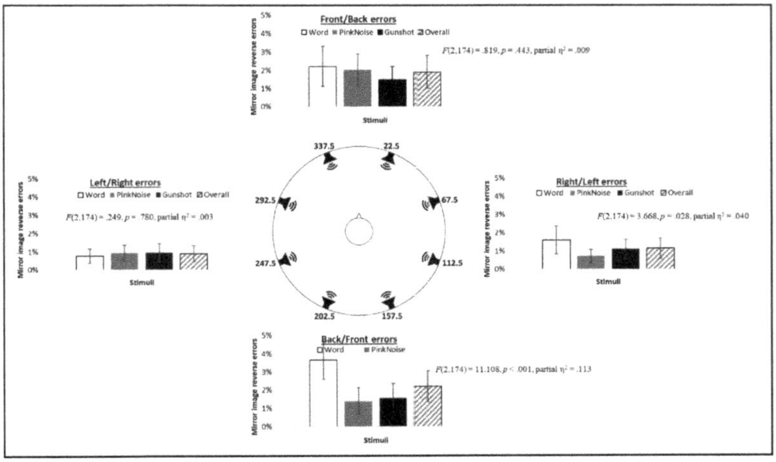

(**b**) Stimuli by monitor angle (no-HPD)

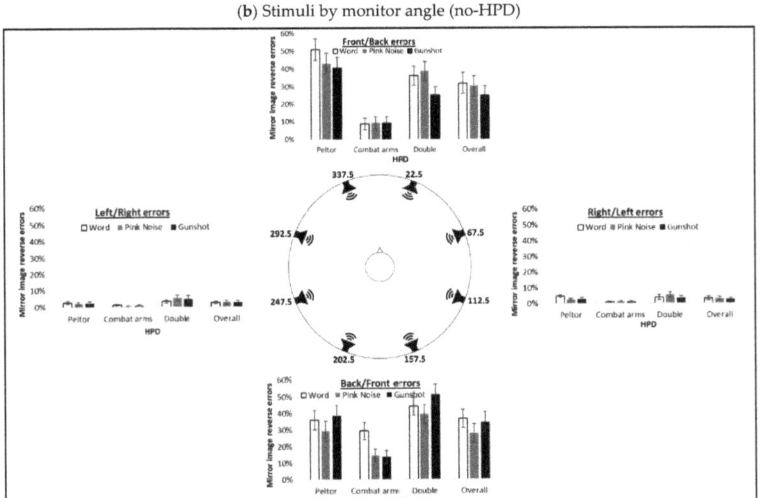

(**d**) Stimuli and HPD by monitor angle

Figure 6. Mean percent of mirror image reverse errors for (**a**) stimuli only (no-HPD condition), averaged across all 90 participants; (**b**) stimuli only (no-HPD condition) at each hemifield, averaged across all 90 participants, including statistics for main effect for stimuli in each hemifield; (**c**) stimuli only (with HPD condition), averaged across HPDs, for all 90 participants; and (**d**) stimuli and HPD at each hemifield, averaged across 30 participants in each HPD group. * $p < 0.05$.

Table 5. Post hoc LSD analyses between hemifields under (a) no-HPD and (b) HPD conditions.

(a) no-HPD				
	L/R	**F/B**	**B/F**	**R/L**
	L/R	0.010	0.013	0.003
(b) with-HPD		**F/B**	0.003	0.008
R/L	**R/L**		**B/F**	0.011
B/F	0.301	**B/F**		**R/L**
F/B	0.263	−0.038	**F/B**	
L/R	0.002	0.303	0.265	**L/R**

Note. Significant comparisons are marked with shaded cells.

A significant Stimulus Type X Hemifield interaction was found ($F(9,783) = 3.774$, $p = 0.001$, partial $\eta^2 = 0.042$). Figure 6b presents the mean percentage of mirror image reversal errors for different stimuli. Separate repeated measures ANOVAs for each hemifield showed significant main effects for stimulus type only for B/F and R/L errors, and not for F/B and L/R errors. In both B/F and R/L hemifields, the word had significantly more B/F and R/L errors than pink noise, and more B/F errors than the gunshot (Table 6a).

Table 6. Comparisons of mirror image reversal errors in each hemifield for the (a) no-HPD condition (LSD analyses between stimuli) and (b) HPD condition (main effects for each HPD, and post-hoc LSD analyses between stimuli for each HPD).

(a) no-HPD	**L/R**	**F/B**	**B/F**	**R/L**
Word vs. pink noise [1]	−0.002	0.002	0.023	0.009
Word vs. gunshot [1]	−0.002	0.007	0.021	0.005
Pink noise vs. gunshot [1]	−0.001	0.006	−0.002	−0.004

(b) with-HPD	**L/R**	**F/B**	**B/F**	**R/L**
Peltor™ [2]	0.027, 0.001	4.234, 0.127	3.830, 0.117	0.514, 0.017
Word vs. pink noise [1]	0.001	0.080	0.065	0.008
Word vs. gunshot [1]	0.001	0.103	−0.027	0.006
Pink noise vs. gunshot [1]	0.007	0.023	−0.092	−0.002
Combat Arms™ [2]	0.635, 0.021	0.099, 0.003	11.387, 0.282	0.125, 0.004
Word vs. pink noise [1]	0.003	−0.007	0.148	0.000
Word vs. gunshot [1]	0.002	−0.005	0.157	−0.002
Pink noise vs. gunshot [1]	−0.002	0.002	0.008	−0.002
Double protection [2]	0.329, 0.011	6.415, 0.181	4.258, 0.128	3.083, 0.096
Word vs. pink noise [1]	−0.009	−0.027	0.047	−0.013
Word vs. gunshot [1]	−0.003	0.110	−0.072	0.009
Pink noise vs. gunshot [1]	0.007	0.137	−0.118	0.022

[1] LSD test; [2] F(2,58), partial η^2

Note. Significant comparisons are marked with shaded cells.

3.2.2. With-HPD

Main effects were found for stimulus type, hemifield, and HPD type ($F(2,174) = 11.589$, $p < 0.001$, partial $\eta^2 = 0.118$; $F(3,174) = 58.829$, $p < 0.001$, partial $\eta^2 = 0.403$; and $F(1,87) = 581.961$, $p < 0.001$, partial $\eta^2 = 0.870$, respectively). As with the no-HPD condition, the word had the largest percentage of mirror image reversal errors (Figure 6c). There were also more F/B and B/F reversal errors than R/L and L/F errors (Table 5b). The Combat Arms™ 4.1 had less errors than both the Peltor™ H515FB and double-protection HPDs did (LSD = −0.138, $p < 0.001$ and LSD = −0.142, $p < 0.001$, respectively).

Significant interactions were found for Stimuli Type X HPD Type, Hemifield X HPD Type, Stimuli Type X Hemifield, and Stimuli X Hemifield X HPD Type ($F(4,174) = 3.703$, $p = 0.006$, partial $\eta^2 = 0.078$; $F(6,261) = 5.965$, $p < 0.001$, partial $\eta^2 = 0.121$; $F(6,522) = 6.909$, $p < 0.001$, partial $\eta^2 = 0.074$; and $F(14,522) = 4.568$, $p < 0.001$, partial $\eta^2 = 0.095$). Figure 6d presents the mean percent of mirror image reversal errors by each HPD for the different stimulus types. Separate repeated measures ANOVAs showed main effects for stimuli perceived while using different HPDs only for F/B and B/F errors, with main effects for stimuli perceived with all HPDs for B/F errors, and only with Peltor™ and double protection for F/B errors (Table 6b). The word stimulus had the largest percentage of errors: more F/B errors than for the gunshot when using the Peltor™ or double-protection HPDs, and more B/F errors than both pink noise and gunshot when using the Combat Arms™ HPD. Pink noise displayed more F/B hemifield errors than the gunshot stimulus when using double-HPD protection, but less B/F errors than the gunshot when using the Peltor™ or double-protection HPDs (Table 6b).

4. Discussion

In the present study, we tested word localization, compared with pink noise and gunshot noise, considering the usage of various HPDs and the source angle/hemifield. Localization of a word was more difficult than the other stimuli. This finding was robust and consistent under conditions of HPD or no-HPD use, and when data were analyzed for the mean RMSE and for mirror image reversal errors. Evident as well was a difference in the effect of the sound source angle and hemifield on localization, depending on whether the listener used, or did not use, HPDs. With no HPDs used, localization was more accurate for sounds delivered from the front. However, with HPDs, the mean RMSE and mirror image reversal errors for stimuli delivered from the front increased, and localization was more accurate for stimuli delivered from the left or right. These findings demonstrate the sensitivity of localization to different characteristics of the environment, listener, and stimuli.

It is well-established that a stimulus' spectrum, i.e., the energy content within the frequency bandwidth, affects the ability to localize it [3,17,19,23,30]. The stimuli used in this study—a word (/esh/), pink noise, and a gunshot—differed in their bandwidth and frequency region. Pink noise and the M16 gunshot (recorded at 200 feet (60.96 m) from the source) had wider bandwidths than the word did, and had higher energy levels in the low and high frequencies, as opposed to the word whose high energy level was mainly in middle-range frequencies. In addition, the accumulated energy of the gunshot and pink noise was much higher than that of the word, with the gunshot having 50% more accumulated energy than pink noise did, and the pink noise having about 50% more accumulated energy than the word did. Accordingly, the pink noise and gunshot were more accurately localized under both no-HPD and with-HPD conditions, with the gunshot being better localized than the pink noise.

The findings of the few previous studies that tested localization of speech stimuli among normal hearing participants support our findings. Jones et al. [11] tested both pink noise and monosyllabic consonant-nucleus-consonant words, similar to the word /esh/ that was used in the present study. However, their normal-hearing participants were tested in a *virtual* acoustic space, namely, stimuli were treated to simulate being delivered from points around the participant and were delivered through headphones. Studies show that localization in a virtual acoustic space is more difficult than in a real one [11,36]. Nevertheless, in line with our findings, they also showed that word localization was more difficult than pink noise, as evidenced by a smaller mean RMSE than for that of words. Other studies compared speech signals to white noise and showed that speech stimuli were localized more poorly than white noise [23–25]. Borg et al. [7] showed that gunshot noise was localized much better than a dog bark, which had a much narrower spectrum with almost no energy in high frequencies. Although a different speech content was tested, these studies, along with the present, show that speech stimuli are more difficult to localize than other stimuli.

A greater difficulty in localizing the word, in comparison with the pink noise and gunshot, was also evident when examining the sound source angle/hemifield. When the sound source was difficult to localize, such as from the back, or from the front when HPDs were used, the difficulty in localizing the word (relative to the pink noise and gunshot) was enhanced. These results show that, in spite of the interactions shown between stimulus type and source angle, and also with HPD use, the pattern of localizing different stimuli was maintained and changed only in its magnitude. Hence, these data indicate differential sensitivity to localizing varying stimuli at different angles. This point should be further studied, mainly in relation to the availability of different localization cues in different source locations.

In terms of localization, comparison between HPDs showed that, as expected, performance worsened with HPD use. The mean RMSE (across all stimuli, angles, hemifields, and HPDs) increased from 23° to 50°, and the mean mirror image reverse error increased from 2% to 16%, with HPD use. Nevertheless, the use of Combat Arms™ earplugs resulted in much smaller and fewer errors than those with Peltor™ earmuffs and double protection, especially for sounds originating from the front monitors. Indeed, while the mean RMSEs were smallest for the front monitors when HPDs were not used, when HPDs were used, the mean RMSEs for the front monitors exceeded the mean RMSEs for the right and left monitors. Previous studies also found better localization with earplugs than earmuffs [4,7,8,34,37].

Some studies attributed the difficulty in localization mainly to the attenuation level [8,26], while others attributed it to the limitation of pinnae cues [4,6]. The present study provides support to both lines of evidence. The general decline in localization accuracy when all types of HPDs were used is evidence for the effect of attenuation on localization. However, with HPD use, localization errors for sounds coming from the front and back increased more than sounds coming from the right or left. This serves as evidence for the effect of pinnae cues, or their limitation, on localization. In general, the localization of sounds from the right and left is achieved using interaural differences in time and level, while the localization of sounds from the front and back is achieved primarily using cues related to the pinnae, head, and torso. A larger increase in localization errors to sounds coming from the front and back, than from the right and left, implies a greater reduction in pinnae cues, rather than interaural cues.

In the present study, we analyzed the data in two ways that demonstrated participant ability, or, more accurately, inability—as both measures focused on errors—to localize various sounds under different conditions. The measure of mean RMSE is a classic way of analyzing localization and indicating precision level. The measure of mirror image reverse error demonstrates when the listener misses the appropriate localization cues. In real world scenarios, a mirror image reverse error reflects when a listener turns their head or body in the wrong direction; therefore, this measure can be considered more ecological than the mean RMSE. Brown [13] used a measure of Very Large Errors (VLEs), defined as errors larger than 45°; their purpose was also to present localization data ecologically by showing the conditions in which listeners will turn their head in the wrong direction. Other studies used the measure of mirror image reverse errors [7–9]. These measures can be very useful when predictions are made about listener behavior in real world situations.

Along these lines, the present study tested localization in a 360° array. Some previous researchers chose to study localization in only a partial array range, testing stimuli from the front or the side only [3,14,17,30,38,39]. This enables a focus on designated azimuths and selected cues. In the present study, however, we chose to study localization in a circular array to provide information on all cues and learn about listener behavior in a real-world situational design in which sound can be delivered from all directions around the listener. Thus, the current study's data provide ecological predictions on localization.

The data provided by the current study have implications for both listeners who use HPDs, and those who do not. For both groups, it was difficult to localize the spoken word relative to the other stimuli. Localizing speech is especially critical in potentially dangerous situations, in any outdoor activity, at work, and in military training and combat, when erroneous localization of a single short word might be crucial. Speech localization is also critically important to aging adults whose hearing sensitivity deteriorates, or other hearing-impaired listeners. Moreover, for those who use HPDs, the importance of studying the localization of speech is demonstrated by the findings of Yehudai et al. [40] on the necessity of removing HPDs to improve communication in battle due to the difficulty of perceiving speech sounds (e.g., commands) while using HPDs. That said, for those who need to localize speech in loud environments, removing their HPDs can endanger their hearing. The current study's findings suggest that when localization is important, earplugs should be preferred over earmuffs. A similar recommendation was previously suggested by Noble and Russell [4].

While the current study's findings are in line with previous studies, they are applicable to the particular stimuli and HPDs that we tested. Therefore, exploration should be extended to other stimuli and HPDs. Aside from testing additional types of HPDs with different attenuation spectra and levels, testing different types of speech stimuli is also necessary, particularly words and sentences with different durations and spectra. In addition, following Derey et al. [37] who showed the effect of the sound category (vocalizations, traffic sounds, or tones) and behavioral relevance (neutral versus fearful sounds with differential effects on behavior) on localization, the effect of the content's relevance and the speaker's familiarity to the listener should also be tested. An additional limitation of the present study is that the attenuation level of HPDs was measured generally, using a free-field microphone, and not individually for each participant; adding the measurement of individual attenuation levels as a predicting factor for localization would be worthwhile.

In sum, the present study focused on the localization of three sounds and demonstrated more difficulty in localizing a monosyllabic word than pink noise and gunshot noise. In addition, the current study findings emphasize the importance of pinna cues in localization by demonstrating the effect of their reduction when using HPDs.

Author Contributions: Conceptualization, L.F. and N.F.; methodology, L.F. and N.F.; designing the experimental software, L.F. and N.F.; formal analysis, L.F. and N.F.; writing, L.F. and N.F.; supervision, L.F. and N.F.; administration, L.F. and N.F.; funding acquisition, L.F. and N.F.; All authors have read and agreed to the published version of the manuscript.

Funding: The research reported in this publication was supported by the Israel Defense Forces (IDF) Medical Corps and Directorate of Defense Research & Development, Israel Ministry of Defense (IMOD DDR&D) grants #4440796070.

Institutional Review Board Statement: The study was conducted according to the guidelines of the Declaration of Helsinki, and approved by the Institutional Review Board of Ariel University (approval #AU-LF-20181122, 22 November 2018)

Informed Consent Statement: Written informed consent was obtained from all subjects involved in the study.

Data Availability Statement: The data presented in this study are available on request from the corresponding author. The data are not publicly available, due to ongoing analysis for the data.

Acknowledgments: We thank Hadar Hassid for assistance in conducting the experiment.

Conflicts of Interest: The authors declare no conflict of interest.

References

1. Gelfand, S.A. *Hearing: An Introduction to Psychological and Physiological Acoustics*; CRC Press: New York, NY, USA, 2017.
2. Litovsky, R. Development of the Auditory System. In *Handbook of Clinical Neurology*; Elsevier: Waltham, MA, USA, 2015; Volume 129, pp. 55–72.

3. Yost, W.A.; Zhong, X. Sound source localization identification accuracy: Bandwidth dependencies. *J. Acoust. Soc. Am.* **2014**, *136*, 2737–2746. [CrossRef] [PubMed]
4. Noble, W.G.; Russell, G. Theoretical and practical implications of the effects of hearing protection devices on localization ability. *Acta Otolaryngol.* **1972**, *74*, 29–36. [CrossRef]
5. Abel, S.M.; Tsang, S.; Boyne, S. Sound localization with communications headsets: Comparison of passive and active systems. *Noise Health* **2007**, *9*, 101–108. [CrossRef]
6. Abel, S.M.; Boyne, S.; Roesler-Mulroney, H. Sound localization with an army helmet worn in combination with an in-ear advanced communications system. *Noise Health* **2009**, *11*, 199–205. [CrossRef]
7. Borg, E.; Bergkvist, C.; Bagger-Sjöbäck, D. Effect on directional hearing in hunters using amplifying (level dependent) hearing protectors. *Otol. Neurotol.* **2008**, *29*, 579–585. [CrossRef] [PubMed]
8. Brungart, D.S.; Kordik, A.J.; Simpson, B.D.; McKinley, R.L. Auditory localization in the horizontal plane with single and double hearing protection. *Aviat. Space Environ. Med.* **2003**, *74*, 937–946. [CrossRef]
9. Vause, N.L.; Grantham, D.W. Effects of earplugs and protective headgear on auditory localization ability in the horizontal plane. *Hum. Factors* **1999**, *41*, 282–294. [CrossRef] [PubMed]
10. Dorman, M.F.; Natale, S.; Loiselle, L. Speech understanding and sound source localization by cochlear implant listeners using a pinna-effect imitating microphone and an adaptive beamformer. *J. Am. Acad. Audiol.* **2018**, *29*, 197–205. [CrossRef] [PubMed]
11. Jones, H.; Kan, A.; Litovsky, R.Y. Comparing sound localization deficits in bilateral cochlear-implant users and vocoder simulations with normal-hearing listeners. *Trends Hear.* **2014**, *18*, 1–16. [CrossRef]
12. Potts, L.G.; Skinner, M.W.; Litovsky, R.A.; Strube, M.J.; Kuk, F. Recognition and localization of speech by adult cochlear implant recipients wearing a digital hearing aid in the nonimplanted ear (bimodal hearing). *J. Am. Acad. Audiol.* **2009**, *20*, 353–373. [CrossRef]
13. Brown, A.D.; Beemer, B.T.; Greene, N.T.; Argo IV, T.; Meegan, G.D.; Tollin, D.J. Effects of active and passive hearing protection devices on sound source localization, speech recognition, and tone detection. *PLoS ONE* **2015**, *10*, e0136568. [CrossRef]
14. Brungart, D.S.; Kordik, A.J.; Simpson, B.D. The effects of single and double hearing protection on the localization and segregation of spatially-separated speech signals (L). *J. Acoust. Soc. Am.* **2004**, *116*, 1897–1900. [CrossRef]
15. Joubaud, T.; Zimpfer, V.; Garcia, A.; Langrenne, C. Sound localization models as evaluation tools for tactical communication and protective systems. *J. Acoust. Soc. Am.* **2017**, *141*, 2637–2649. [CrossRef]
16. Fink, N.; Pikkel, H.Z.; Eisenkraft, A.; Banta, G.A. Hearing protection devices and methods used for their evaluation: A military perspective. *J. Mil. Veteran Fam. Health* **2019**, *5*, 141–147. [CrossRef]
17. Yost, W.A. Sound source localization identification accuracy: Level and duration dependencies. *J. Acoust. Soc. Am.* **2016**, *140*, EL14–EL19. [CrossRef] [PubMed]
18. Middlebrooks, J.C. Auditory space processing: Here, there or everywhere? *Nat. Neurosci.* **2002**, *5*, 824–826. [CrossRef] [PubMed]
19. Stevens, S.S.; Newman, E.B. The localization of actual sources of sound. *Am. J. Psychol.* **1936**, *48*, 297–306. [CrossRef]
20. Ihlefeld, A.; Shinn-Cunningham, B.G. Effect of source spectrum on sound localization in an everyday reverberant room. *J. Acoust. Soc. Am.* **2011**, *130*, 324–333. [CrossRef]
21. Middlebrooks, J.C. Narrow-band sound localization related to external ear acoustics. *J. Acoust. Soc. Am.* **1992**, *92*, 2607–2624. [CrossRef]
22. Blauert, J. *Spatial Hearing: The Psychophysics of Human Sound Localization*; MIT Press: Cambridge, MA, USA, 1997.
23. Chau, W.; Duda, R.O. Combined monaural and binaural localization of sound sources. *Proc. Asilomar.* **1996**, *2*, 1281–1285.
24. Davis, R.J.; Stephens, S.D.G. The effect of intensity on the localization of different acoustical stimuli in the vertical plane. *J. Sound Vib.* **1974**, *35*, 223–229. [CrossRef]
25. Shigeno, S.; Oyama, T. Localization of speech and non-speech sounds. *Jpn. Psychol. Res.* **1983**, *25*, 112–117. [CrossRef]
26. Simpson, B.D.; Bolia, R.S.; McKinley, R.L.; Brungart, D.S. The impact of hearing protection on sound localization and orienting behavior. *Hum. Factors* **2005**, *47*, 188–198. [CrossRef] [PubMed]
27. Dobreva, M.S.; O'Neill, W.E.; Paige, G.D. Influence of aging on human sound localization. *J. Neurophysiol.* **2011**, *105*, 2471–2486. [CrossRef]
28. Gordon, M.S.; Russo, F.A.; MacDonald, E. Spectral information for detection of acoustic time to arrival. *Atten. Percept. Psychophys.* **2013**, *75*, 738–750. [CrossRef]
29. Rychtáriková, M.; Chmelik, V.; Roozen, N.B.; Glorieux, C. Front–back localization in simulated rectangular rooms. *Appl. Acoust.* **2015**, *90*, 143–152. [CrossRef]
30. Yost, W.A.; Loiselle, L.; Dorman, M.; Burns, J.; Brown, C.A. Sound source localization of filtered noises by listeners with normal hearing: A statistical analysis. *J. Acoust. Soc. Am.* **2013**, *133*, 2876–2882. [CrossRef] [PubMed]
31. Abel, S.; Burrell, C.; Saunders, D. *The Effects of Integrated Hearing Protection Surround Levels on Sound Localization, Scientific Report*; DRDC-RDDC2015-R012; Defense Research and Development Canada: Ottawa, ON, Canada, 2015.
32. Alali, K.; Casali, J.G. Auditory backup alarms: Distance-at-first-detection via in-situ experimentation on alarm design and hearing protection effects. *Work* **2012**, *41* (Suppl. 1), 3599–3607. [CrossRef]
33. Lee, K.; Casali, J.G. Development of an auditory situation awareness test battery for advanced hearing protectors and TCAPS: Detection subtest of DRILCOM (detection-recognition/identification-localization-communication). *Int. J. Audiol.* **2017**, *56* (Suppl. 1), 22–33. [CrossRef]

34. Talcott, K.A.; Casali, J.G.; Keady, J.P.; Killion, M.C. Azimuthal auditory localization of gunshots in a realistic field environment: Effects of open-ear versus hearing protection-enhancement devices (HPEDs), military vehicle noise, and hearing impairment. *Int. J. Audiol.* **2012**, *51* (Suppl. 1), S20–S30. [CrossRef]
35. Sintonizar Productions and TheWhiteKnuckleFilms. Real life Gunshot Sound Effects. Gunshot Sound Effect. ASMR. YouTube. Available online: https://www.youtube.com/watch?v=1N_m3tsPyP0 (accessed on 10 October 2021).
36. Middlebrooks, J.C. Individual differences in external-ear transfer functions reduced by scaling in frequency. *J. Acoust. Soc. Am.* **1999**, *106*, 1480–1492. [CrossRef] [PubMed]
37. Laroche, C.; Giguère, C.; Vaillancourt, V.; Marleau, C.; Cadieux, M.F.; Laprise-Girard, K.; Gula, E.; Carroll, V.; Bibeau, M.; Nélisse, H. Effect of hearing and head protection on the localization of tonal and broadband reverse alarms. *Hum. Factors* **2021**. [CrossRef] [PubMed]
38. Derey, K.; Rauschecker, J.P.; Formisano, E.; Valente, G.; de Gelder, B. Localization of complex sounds is modulated by behavioral relevance and sound category. *J. Acoust. Soc. Am.* **2017**, *142*, 1757–1773. [CrossRef] [PubMed]
39. Yost, W.A. Sound source localization identification accuracy: Envelope dependencies. *J. Acoust. Soc. Am.* **2017**, *142*, 173–185. [CrossRef]
40. Yehudai, N.; Fink, N.; Shpriz, M.; Marom, T. Acute acoustic trauma among soldiers during an intense combat. *J. Am. Acad. Audiol.* **2017**, *28*, 436–443. [CrossRef] [PubMed]

MDPI AG
Grosspeteranlage 5
4052 Basel
Switzerland
Tel.: +41 61 683 77 34

Sensors Editorial Office
E-mail: sensors@mdpi.com
www.mdpi.com/journal/sensors

www.ingramcontent.com/pod-product-compliance
Lightning Source LLC
Chambersburg PA
CBHW061624210926
51850CB00009B/2432